# EATING



Peter Singer is the author or editor of over thirty books including *Writings on an Ethical Life*, a comprehensive volume of his best and most provocative writing, *Animal Liberation*, and most recently, *The President of Good and Evil*. Singer is currently the Ira W. DeCamp Professor of Bioethics at Princeton University's Center for Human Values. He lives in New York City.

Jim Mason is an author, speaker, journalist, environmentalist, and attorney who focuses on human and animal concerns. He is the co-author, with Peter Singer, of *Animal Factories*, and his writing has appeared in a variety of publications including the *New York Times*, the *New Scientist* and *Audubon*.

PETER SINGER AND JIM MASON

# EATING

arrow books

Published in the United Kingdom by Arrow Books, 2006

3 5 7 9 10 8 6 4 2



First published under the title *The way we eat: Why our food choices matter* in the United States in 2006 by Rodale Inc.

Arrow Books
The Random House Group Limited
20 Vauxhall Bridge Road, London, SW1V 2SA

Random House Australia (Pty) Limited
20 Alfred Street, Milsons Point, Sydney
New South Wales 2061, Australia

Random House New Zealand Limited
18 Poland Road, Glenfield
Auckland 10, New Zealand

Random House (Pty) Limited
Isle of Houghton, Corner of Boundary Road & Carse O'Gowrie
Houghton 2198, South Africa

Random House Publishers India Private Limited
301 World Trade Tower, Hotel Intercontinental Grand Complex
Barakhamba Lane, New Delhi 110 001,India

Random House Group Limited Reg. No. 954009

www.randomhouse.co.uk

A CIP catalogue record for this book
is available from the British Library

Papers used by Random House
are natural, recyclable products made from wood grown in
sustainable forests. The manufacturing processes conform to
the environmental regulations of the country of origin

ISBN 9780099504023 (from Jan 2007)
ISBN 0099504022

Printed and bound in Great Britain by
Cox & Wyman Ltd, Reading, Berkshire

# CONTENTS

# PREFACE

British food has come a long way—in many different directions—since I arrived in Oxford from Australia in 1969 to do a postgraduate degree in philosophy. The standard meal then was two slices of roast beef covered in thick brown gravy, with mashed potatoes and overcooked peas. Today, British cooking is far more creative and can draw on a wider range of fresh vegetables and fruits. But I have in mind changes that go far beyond our enjoyment of food. It was during my student years in Oxford that I learnt, with a shock, that my eggs did not come from hens scratching around in a farmyard, but from birds who spent their entire productive lives crammed into wire cages too small for them to stretch their wings. My breakfast bacon, I discovered, came from pigs reared indoors in crowded pens on bare concrete floors. Worse still, the mothers of these pigs had spent their pregnancies in stalls too narrow to turn around in. In those days, I even ate veal—until I read that it came from calves housed in narrow individual stalls, deliberately made anaemic to keep their flesh the pale pink colour for which gourmet restaurants were prepared to pay the highest price. I began to think about the ethics of how we should treat animals, and of what I was eating.

Ecological issues were just starting to attract serious attention—the world's first environmental summit was held in Stockholm in 1972. That led many people to think about farming methods, but if you wanted organically grown vegetables, your best option was to grow them yourself. Local farmers' markets didn't exist, and organically grown produce was only available in a few small health food stores. They were expensive, and few people sought them out, so what there was sold slowly and usually looked tired and unappetizing.

Today, keeping sows and veal calves in individual stalls is illegal in

Britain and these confinement systems are in the process of being phased out across the European Union. (The fate of a proposed EU ban on the standard battery cages for laying hens hangs in the balance, as I write, pending a review.) Organic foods—not only fresh fruit and vegetables, but a wide range of packaged and processed foods as well—are available in all the major supermarket chains. In some of them, you won't find eggs from caged hens anymore. The "fair trade" label did not exist when I lived in Britain. It can now be found on an increasingly wide range of products, and sales are growing at an extraordinary rate: in 2004, they were 50 percent higher than the previous year. A Market and Opinion Research International survey taken in 2005 showed that every second person in Britain recognizes the fair trade logo, and associates it correctly with the idea of a better deal for producers in developing countries.

These are all important changes for the better. But there is still much to do. Individual sow stalls may have gone, but most pigs are still reared intensively, kept indoors for their entire lives. Almost all chickens and turkeys are still kept in sheds crowded with 10 or 20,000 birds, much like the American production system described elsewhere in this book. Despite the rapid growth of organic food sales, most food is still produced by conventional methods that are not sustainable.

The pages that follow trace the food choices of three families, each with a distinct way of eating. Although the families are American, the ethical implications of their choices need only minor modifications to be applicable to the choices made by British families. Where Britain—and more broadly, Europe—differs from the United States, is in the greater consumer awareness of some of the issues this book discusses, and the readiness of these consumers to change their purchasing habits accordingly. Although factory farming in the U.S. is on an even larger scale than in Britain, most Americans know little about it. The American campaign for fair trade products has a long way to go before it is conceivable that every second American will recognize the label and understand what it signifies. Virtually all Americans are now eating at least some genetically modified foods, but there has been little debate about whether this is a good or a bad thing.

European agricultural policies have implications that reach far beyond Europe. As Asian nations become more prosperous, they are transforming their traditional low-meat diet and their ancient, small-scale, environmentally sustainable methods of producing food. China, for example, now seems to look to American for a model of how to produce the animal products that its large new middle class wants to buy. This could place demands on the environment, not only locally but globally, that far exceed in scale anything we have seen in Europe. The task for Europeans, therefore, is to provide an example of an alternative way of producing food. Europe must demonstrate that it is possible for a major region of the world to grapple with the ethical issues raised by what we eat in a manner that is environmentally sustainable, shows respect for animals, and is fair to all those involved in the food industry. All it takes is enough informed, ethically concerned consumers.

**—Peter Singer**

# EATING

INTRODUCTION

# FOOD AND ETHICS

We don't usually think of what we eat as a matter of ethics. Stealing, lying, hurting people—these acts are obviously relevant to our moral character. So too, most people would say, is our involvement in community activities, our generosity to others in need, and—especially—our sex life. But eating—an activity that is even more essential than sex, and in which everyone participates—is generally seen quite differently. Try to think of a politician whose prospects have been damaged by revelations about what he or she eats.

It was not always so. Many indigenous hunter-gatherers have elaborate codes about who may kill which animals, and when. Some have rituals in which they ask forgiveness of the animals for killing them. In ancient Greece and Rome, ethical choices about food were considered at least as significant as ethical choices about sex.[1] Temperance and self-restraint in diet, as elsewhere in life, were seen as virtues. Socrates, in Plato's *Republic*, advocates a simple diet of bread, cheese, vegetables, and olives, with figs for desert, and wine in moderation.[2] In traditional Jewish, Islamic, Hindu, and Buddhist ethics, discussions of what should and should not be eaten occupy a prominent place. In the Christian era, however, less attention was paid to what we eat—the major concern being to avoid gluttony, which, according to Catholic teaching, is one of the seven cardinal sins.

The way food is sold and advertised doesn't help. Despite the recent upsurge of farmers' markets, in the developed world almost all food is purchased from supermarkets. Shoppers are not presented with relevant information about the ethical choices that surround food. Instead, the world food industry spends more than $40 billion annually trying to make us eat their products—a figure greater than the gross domestic product of 70 percent of the world's nations.[3] That buys an avalanche of advertising that sweeps down on us from all sides but tells us only what the advertisers want us to know. Marion Nestle, a nutritionist who worked in the U.S. Department of Agriculture and on the *Surgeon General's Report on Nutrition and Health* (1988), has described how the food industry has crossed ethical lines in bringing political pressure to bear on what should be dispassionate scientific government advice on how Americans can eat a healthy diet.[4] Morgan Spurlock's *Supersize Me* raised serious ethical questions about the contribution of fast food chains like McDonald's to America's epidemic of obesity.[5] Our focus is not on these issues. There is already plenty of information out there about them. If you enjoy unhealthy food so much that you are prepared to accept the risk of disease and premature death, then, like a decision to smoke or climb Himalayan peaks, that is primarily your own business. Our focus is on the impact of your food choices on others.

## A NEW AWARENESS

Over the last thirty years we've seen the first stirrings of a different kind of concern about what we eat. In Western Europe and the United States, where veal is intensively reared, many people have stopped eating it after learning that veal calves are separated from their mothers soon after birth, deliberately made anemic, denied roughage or the possibility of exercise, and kept in stalls so narrow they cannot turn around. In the United States, veal consumption has fallen to less than a quarter of what it was in 1975.[6] Consumers also increasingly seek out organically produced food, for reasons that range from an ethical concern for the environment to a desire to avoid ingesting pesticides and the conviction that organic food tastes better than food from conventional sources. Today, organic foods can easily be found in supermarkets and are the fastest growing section of the food industry.[7]

Buying organic isn't enough, however, for the millions of vegetarians all over the world who refuse to eat any meat or fish. In the United States, a 2003 Harris poll found that almost 3 percent of the population says they never eat meat, poultry, fish, or other seafood. In Britain, various surveys have indicated that between 5 and 7.6 percent of the population consider themselves vegetarian or vegan. In one survey, 3 percent said that they were vegan.[8] Avoiding meat and fish used to be as far as anyone went. Now vegans, who eat no animal products at all, are as common as vegetarians once were. In fact, the same United States Harris poll found that half of those who said they never eat meat, poultry, fish, or other seafood also said they never eat dairy products, eggs, or honey. And it's not just the vegans who are conscious of food. Throughout developed countries, people are learning to ask tough questions about where their food comes from and how it was produced. Is the food grown without pesticides or herbicides? Are the farm workers paid a living wage? Do the animals involved suffer needlessly?

Questions like these are part of a growing movement toward ethical food consumption. In 2005 two major U.S. supermarket chains, Whole Foods Market and Wild Oats, announced that they would not sell eggs from caged hens, and Trader Joe's said it would not use caged eggs for its own brand of eggs. As John Mackey, Whole Foods Market's CEO, has said, these changes were the result of customer demand.[9] Nor is this concern limited to highly educated people in upper-income brackets. It affects all forms of food consumption, right down to McDonald's and Burger-King, both of which have, as we shall see, recently taken steps that show them to be sensitive to ethical criticism of their products.

Virtually anyone, irrespective of income, can make a positive contribution to this movement. Making better food choices doesn't require hours spent reading labels or rigid adherence to any particular diet. All it takes is the information we provide in this book, which we hope will bring a little more awareness about the significance of the food choices we all make.

## VOTING AT THE SUPERMARKET

Increasingly, people are regarding their food choices as a form of political action. One of the conscientious consumers we interviewed for this book said, "I try to vote with my dollar and not enrich those who are doing bad things in the world."[10]

In Europe, ethical consumption has gone much further than in the

United States. Since the 1980s, non-government organizations have been campaigning to persuade supermarkets to stock products that are fairly traded, free of genetically modified organisms (GMOs), and, in the case of animal products, from producers who avoid the most restrictive forms of confinement. Most major supermarkets in Europe carry free-range eggs, a wide variety of GMO-free products and fairly-traded coffee, tea, chocolate, and bananas. McDonald's introduced organic milk to its British restaurants in 2003.[11] The Co-Op, a national British supermarket chain, now buys its house brand of chocolate from growers who meet fair trade standards. Because the cocoa growers in Ghana receive a higher price for fair trade cocoa, the price of the Co-Op's house brand chocolate increased. The conservative *Daily Telegraph* predicted that consumers would resist paying higher prices for fair trade products. Instead, sales of Co-Op brand chocolate have doubled, while sales of the other brands of chocolate the store stocks have declined. In 2003 the Co-Op converted its own brand coffee to fair trade and in the next year saw Co-Op brand coffee sales rise 20 percent, while sales of other brands fell 14 percent.[12]

The extent to which British consumers choose ethically when buying food is, by American standards, quite astonishing. In Britain, sales of free-range eggs—that is, eggs that are not only from "cage free" hens, but from hens able to walk outside—have now surpassed in value sales of eggs from caged hens.[13] Since 2002, two major British supermarket chains, Marks and Spencer and Waitrose, have sold *only* free-range whole eggs. In the United States, not even Whole Food Markets or Wild Oats have gone that far, although they cater to more environmentally conscious and affluent consumers than the more mainstream British chains. Marks and Spencer has also eliminated eggs from caged hens from its entire food range, requiring every manufacturer from whom they purchase food products to source their eggs from a list of approved and inspected egg producers. Now Tesco is also phasing out eggs from caged hens, and ASDA, the British Wal-Mart affiliate, does not use eggs from caged hens in its own brand eggs. In Britain, at least a third of laying hens are no longer in cages, and about 15 percent are kept by registered free-range producers.[14] In the United States, in contrast, 98 percent of eggs are still from caged hens, and, as we shall see, of the remaining 2 percent, very few of these are truly free-range.[15]

Given the strong British concern for ethical consumption, it is hardly surprising that Rowan Williams, Archbishop of Canterbury, became the first major Christian leader to affirm that ethics extends to food choices. Under his leadership the Church of England has issued a report entitled *Sharing God's Planet* that recommends sustainable consumption and says every Christian has a duty to "care for every part of God's creation." The Church recommends that clergy themselves make eco-friendly consumption choices, selling fairly traded products at church fêtes and using organic bread and wine for communion services.[16]

## THREE FAMILIES

The issues raised by our food choices are clearly illustrated by three families we're about to meet. The families are all American, but they illustrate three different approaches to food that can now be found anywhere in the developed world. American ways of eating, marketing and producing food have a powerful influence on other countries. We'll start with the Hillard-Nierstheimer family, who live in Mabelvale, Arkansas: Lee Nierstheimer; his wife Jake; and their two children, Katie and Max. Their food choices exemplify the Standard American Diet. Jake, who does the family shopping, generally goes to her local Wal-Mart Supercenter because it is hard to beat their prices, and she can get everything in one stop. When they want to go out to eat, the family picks one of the many fast food chains in the area.

Halfway across the country, in Fairfield, Connecticut, we'll sit down to dinner with the Masarech-Motavalli family: Jim Motavalli; his wife, Mary Ann Masarech; and their daughters Maya and Delia. Jim and Mary Ann are concerned about their family's health and about the impact their food purchases have on the environment. Much of the food they buy is organically produced, so they know it is relatively free of pesticides and has not been grown with synthetic fertilizers. In summer and fall Mary Ann likes to go to a local farm to get fresh, locally grown vegetables. But Jim and Mary lead busy lives, and convenience is a factor too, so their purchases don't always quite match up to their ideals.

And in Olathe, Kansas, an outer suburb of Kansas City, we'll talk with the Farb family: Joe; his wife JoAnn; and their daughters Sarina and Samantha. Of our three families, the Farbs follow the strictest ethical principles. Theirs is a vegan household; everything they eat is purely plant-based, and nothing comes from an animal. The Farbs also seek out organically grown food whenever possible.

In getting to know our families, we come to appreciate the individual circumstances in which each of us chooses what to buy and what to eat and the complex personal, social, and economic factors that go into these decisions. As we have already said, we think that these choices have ethical significance, and we will later criticize some of the food choices made by our families. Obviously, though, food choices are only one aspect of what people do and not a sufficient basis for judging their moral character. Indeed, since food ethics has been such a neglected topic in our culture, it is quite likely that otherwise good people are making bad choices in this area simply because they have not really focused on it, or do not have access to the information they need to make good choices.

## KNOWLEDGE IS POWER

Our plan was to note the foods that our families chose and then trace them back through the production process to see what ethical issues arose. Once we found out what our three families ate, we wrote to 87 corporations who had manufactured at least one product that a family had bought. We informed each corporation of our project and asked for their assistance in identifying and facilitating our visits to the farm or facility from which the product came. Few companies bothered to reply. So we sent follow-up letters, adding that we were keen to get the producers' side of the story. After all this, only 14 companies indicated that they were willing to assist us in any way. Most of these companies were relatively small producers of organic foods.

We were disappointed but not surprised. As recently as the 1970s, the food industry was proud to show its farming practices to the public. No more. Not long ago, the producer of an Australian current affairs program suggested doing an interview with Peter in a setting involving animals, somewhere not too far from Princeton, New Jersey. Peter said: "Fine, let's do it inside an intensive farm so that viewers can see where their meals come from." The producer agreed and said he'd find a location. Several days later he called back to admit defeat. He'd contacted several intensive producers and not one of them would let the television cameras in. He had even turned for assistance to the Animal Industry Foundation, headed by Steve Kopperud, probably America's most forceful defender of the animal production industries.

Kopperud travels the country giving speeches at animal industry conferences, telling producers that they must take the offensive against the animal rights movement by communicating with the public and giving consumers accurate information about the way producers treat their animals. In a column in *Florida Agriculture*, Kopperud blamed the media for being out of touch with rural life. "The CBS brass should set the corporate jet down in rural America and take a look around," he wrote. "Follow a farmer around for a season, or have a group of city dwellers try to tackle the hard work of farming or ranching. Now that would be a dose of reality!"[17] But the farmer won't let the CBS brass follow him around—not when he goes inside the factory farm doors, anyway. Kopperud was unable—or unwilling—to help the television producer find a single egg, chicken, veal, or pig operation that would let the cameras in.

We contacted Kopperud again while we were working on this book. The email correspondence went like this:

> January 24, 2005
>
> Hello Steve Kopperud,
>
> Perhaps you remember my name from my book with Peter Singer, *Animal Factories*. We're at work on a new book now that will cover a wide range of ethical concerns that consumers have today about farming and food production. It will include discussions of current concerns about plant agriculture as well, such as labor, environment, fair trade, corporate responsibility, and so on . . . Would you be willing to give us an on-the-record interview for our book? If so, I live just a few hours outside of Washington, D.C., and could meet you to talk in person.
>
> Jim Mason

Kopperud replied promptly, saying that he wasn't sure of his schedule and asking Jim to contact him again in a week or so. Jim did that but got no reply, so he wrote again a week later. Another week passed without reply, so Jim wrote a fourth time. This time Kopperud did reply, but only to say that it was a hectic time of year and he was going out of town, so perhaps it would be better if we sought out "another whose views are similar to mine." Jim replied that we were particularly keen to talk to Kopperud himself and referred to Kopperud's column in *Florida Agriculture* taking the media to task for being out of touch with rural life. There was no response to that message, nor to any of four further reminders sent over the next six weeks.

Around the time we were trying to talk to Kopperud, we read in the farm journal *Feedstuffs* that the National Pork Board was in the process of training more than 200 producers to help them better communicate with neighbors and communities about modern pork production. The article quoted Danita Rodibaugh, a vice president of the National Pork Board and a pork producer, as saying, "One way to tell the industry's story is from a producer armed with the facts. It's difficult to maintain a negative view of an industry if you put a face on it and tell your side of the story when misinformed critics attack it."[18] Great, we thought, here is a pork producer who will be keen to show us how she keeps her pigs. But when we contacted her, she told us that it wouldn't be possible for us to visit her farm, because of concern that we might spread diseases among her pigs. She offered to send us a *Pork Facts* book instead. We declined that offer and instead offered to buy sterile, disposable full-length gowns, overboots, caps, and surgical masks for visiting her farm and to meet, or exceed, whatever biosecurity procedures she required for her employees, visiting veterinarians, or others who she may admit to her farm. We received no response to this message, nor to two further follow-up messages.

We don't take this personally. Journalists looking into how our food is produced have had the door slammed in their faces over and over again. When Moark, which boasts of being the Nation's #1 Egg Marketer, announced plans to build an egg production unit housing 2.6 million hens in Cherokee County, Kansas, local residents protested. Roger McKinney, a staff writer for the local newspaper the *Joplin Globe*, contacted the company and asked to see one of its existing farms. McKinney was granted permission but was not allowed to bring a photographer. According to McKinney, a company official told him that "the company doesn't allow photographs inside the barn because many people would not understand why the birds are in cages."

In St. Louis, Missouri, news channel KSDK-TV ran into the same problem with its coverage of the way in which two of the nation's biggest pork producers, Cargill Pork and Premium Standard Farms, had managed to get pigs raised in Missouri despite that state's laws against corporate farm ownership. KSDK was denied access to any of the pig farms. Spokesmen for both Cargill and Premium Standard declined to go on the record about their activities. The Cargill Pork spokesman said "a television story is not the best way for the company to tell its story." The Missouri Pork Association, an affiliation of pork producers in the state, took three days to return the reporter's telephone calls and then said that the association would be "unlikely" to be able to help.[19] Kevin Murphy, vice-president of Vance Publishing, publisher of *Food Systems Insider*, a monthly magazine for the food industry, has found the same secretive mentality all the way up the chain that leads from producer to consumer: "I can't tell you how many times I've been in rooms where people say, 'Well, our objective is just to be quiet, to just get out of the public eye as quickly as possible.'"[20]

There are rare exceptions. In a pig industry magazine we read about an Iowa pork producer who was quoted as saying that the only way to stop attacks on intensive farming by animal rights organizations was "to get in front of the public and tell them our story: the real story, not their lies." We called him, expecting to be given the usual excuses. After a couple of cautious phone calls, he agreed to allow us to visit. "What about biosecurity?" we asked. "It's not an issue if you're not coming from other farms," he replied. We assured him we wouldn't go near any pigs before visiting his farm, and that was it. He turned out to be a blunt man who didn't see any reason to hide what he was doing. While he showed us the various stages of intensive pig production on his property, he forthrightly defended what he was doing, both in terms of producing pork at a price everyone can afford and in terms of keeping the pigs comfortable.

Within animal agriculture, a few people speak frankly about why the industry is so secretive. Peter Cheeke is a professor of animal science and the author of the widely used textbook *Contemporary Issues in Animal Agriculture*, in which he writes: "For modern animal agriculture, the less the consumer knows about what's happening before the meat hits the plate, the better. . . . One of the best things modern animal agriculture has going for it is that most people in the developed countries are several generations removed from the farm and haven't a clue how animals are raised and processed." Cheeke then gets more specific, stating that if urban meat eaters were to see the raising and processing of industrially produced chickens, "they would not be impressed." Many of them might even "swear off eating chicken and perhaps all meat."[21] Another agriculture professor, Wes Jamison, agrees that "There is a gulf between the reality of animal production and the perception of animal production in the non-farming American public." But Jamison, who teaches at Dordt College in Sioux Center, Iowa, doesn't think that this gulf will ever be bridged. "You're not going to see a beef-packing plant be transparent. They can't. It's so shocking to the average person."[22]

After suggesting that animal agriculture benefits from public ignorance about production methods, Professor Cheeke invites readers of his textbook—who are mostly university agriculture students—to ask themselves a crucial question: "Is this an ethical situation?" Industry officials who would resist contemplating Cheeke's question would do well to heed the counsel of Lord Acton, who said, "Everything secret degenerates . . . nothing is safe that does not show how it can bear discussion and publicity."

Just as the food industry resists disclosing general information about its food production, so it has resisted offering us information about the foods purchased by the three families we monitored. Yet even without the industry's cooperation, we were able to discover a great deal about how these families' foods were produced, and this can help food consumers—that is, all of us—make better, more ethical food choices for ourselves and for our families.

# PART I

# EATING THE STANDARD AMERICAN DIET

# 1

# JAKE AND LEE

There is no downtown, no bustling public square, no quaint historic district in Mabelvale, Arkansas. The "main drag" is Baseline Road—four lanes of traffic running through a corridor of gas stations, convenience stores, and strip malls in the urban sprawl southwest of Little Rock, to which it was annexed in 1980. Sixty percent of Mabelvale's 5,000 inhabitants are white, 25 percent are African-American, and 10 percent are Latino; they live in homes worth around $75,000 and earn about $30,000 annually.

Among the residents of Mabelvale is the family of Jake Hillard, 36, and Lee Niersthelmer, 26. We chose them for their basic meat-and-potatoes diet—sometimes called the Standard American Diet, or SAD. Though the term lacks a precise definition, it is the most widely eaten diet in America. The Standard American Diet is high in meat, eggs, and dairy products. Carbohydrates such as bread, sugar, and rice are usually eaten in refined form, which, combined with a low intake of fruit and vegetables, means that the diet is low in fiber. Frequent consumption of fried foods contributes to a high intake of fat, with as much as 35 percent of calories coming from fat, most of it saturated and much of it animal fat. A burger on a bun with a serving of french fries, followed by an ice cream sundae and washed down with a can of cola, fits squarely in this American tradition. It's a quick and easy way of putting enough food in your stomach to feel satisfied. With America's low prices for meat, eggs and dairy products, it's not expensive either.

We met Lee Niersthelmer at his place of work, a local firm that makes custom-made handling systems and conveyors for major manufacturers. A man of medium height and build, he has a boyish face and a full head of straight brown hair. He tells us that had we come a few months earlier, he would have been at work in the machine shop, welding and bending metal into the sizes and shapes called for in customers' specifications. But he has recently been promoted and is now an engineer, designing and drawing plans for the equipment manufactured by his company. It's the end of the working day and he takes us back to his home, where he lives with his wife, Jake, and their two children, Katie, 2, and Max, 6 months. They are at the end of a dead-end street in a neighborhood of modest homes that date from the 1950s and 1960s. On the corner is a little old house renewed by white vinyl siding, next to a tattered blue mobile home, then a neat, small, brick house, then a couple more clad in vinyl, and so on. At their gate we're greeted by a couple of very friendly dogs: one looks like a mid-sized St. Bernard—large, fluffy, brown-and-white. "That's Baggie," Lee says. The other one, Annie, a Border Collie with maybe a bit of Australian Shepherd mixed in, is the current neighborhood hero—she roused several people in time to catch a burglar in the act of breaking and entering a house up the street.

The yard, walkway, and stoop are cluttered with bright, primary-colored plastic tricycles, wagons, miniature chairs, balls, and toys. Inside, there's more of the same, with Jake—snugly curled up in an overstuffed chair—breast-feeding baby Max. At her feet, Katie is engrossed in watching *Finding Nemo* on the VCR. A black-and-white cat dozes among the toys on the sofa. Lee immediately goes over to Katie, kisses her, then Jake, and takes the baby in his arms.

Jake gets up apologizing for "the mess," saying she's tired from being up all night with Max, who has been fussy with teething and allergies lately. She's nearly Lee's height, with a full, pretty face. She wears her auburn hair long and straight, with a thick hank of bangs that curl down to her eyes. They show us around the house, including the kids' room, which they have painted and decorated. Then it's time for dinner. Jake serves Katie her favorite meal: macaroni and cheese, green beans, and a slice of bread and butter. Katie, between giggles, sips from her glass of milk and takes bites of the macaroni and cheese. Lee adjusts Max's high-chair and then spoon-feeds him bites of pureed spinach lasagna and green beans with potatoes. Meanwhile, Jake puts food on plates for herself and her husband. Tonight they're having barbequed chicken breasts, a lettuce and tomato salad, and some of the green beans that Katie is also eating, but seasoned in the Southern way with bits of bacon and onion.

There is a small plate of paprika-sprinkled deviled eggs on the table, which Lee had snitched from a large platter in the refrigerator while Jake was tending to the chicken under the broiler. Jake takes one, and in a tone more teasing than scolding, tells Lee that she made them for tomorrow's family picnic with her parents. Then she takes a bite, which sends Katie into a fit of giggles.

Lee is drinking a Samuel Adams beer and Jake a Diet Coke.

After dinner, Lee clears the table and rinses the dishes while Jake tends to Max.

"Can we have some ice cream now?" Katie burbles, and, seeing her mother's look, quickly adds, "Please?"

"Only if we have some strawberries too," Jake says.

"And chocolate sauce," chimes in Lee from the sink.

"Oh, brother," Jake says, rolling her eyes. "It's chocolate chip."

Lee takes a tub of ice cream from the freezer and puts it on the table. "The berries are in that white bowl on the bottom shelf,' Jake says, and after two beats she adds, "You've got to be kidding about that chocolate sauce, right?"

## GROCERY SHOPPING AT WAL-MART

The next day, Jake arranges for child care and takes us on her shopping trip to the Wal-Mart Supercenter on Baseline Road. As we enter the store we find the manager and explain our project. When we tell him that we want to use a video camera to tape Jake's shopping trip, he becomes agitated and tells us that this is against company policy. Permission to do so "would have to come from Bentonville," he says, referring to the national head office, and indicates that it is rarely given. After assuring him that we will be leaving the video recorder outside, we ask about using a small pocket audiotape recorder. He becomes even more agitated and emphasizes the company's policy against recording of any kind in its stores. Defeated, we go back to Jake's car, lock up the equipment, gather pens and a notebook, and get on with the shopping.

We begin at the dairy case, where Jake picks up a half-gallon of milk. "Skim milk for momma. Great Value—that's a Wal-Mart brand. I get Coleman Dairy whole milk for Katie; it's kind of a local brand. They're down in Batesville." Next she picks up a carton of a dozen

"Country Creek" eggs. The fine print says: "Moark Productions, Inc., 1100 Blair Avenue, Neosho, MO." There is a logo on it as well—it has the words "Animal Care Certified" in a semi-circle, and there is a big check mark in the middle. Jake moves along the dairy case pulling out products and putting them in the shopping cart: Oscar Mayer bacon, Daisy Sour Cream, Great Value Extra Sharp Cheddar, and Kraft 100 percent parmesan cheese. From the meats, Jake picks out Armour pepperoni, Petit Jean brand peppered bacon, Jimmy Dean sausage, some store brand skinless chicken breasts, an unlabeled package of "beef loin porterhouse steak," Ball Park corn dogs, and Advance Brand "steak fingers." She gets orange juice, too, and some vegetables, including an iceberg lettuce and tomatoes. So it goes, aisle after aisle, until we have enough of the favorite foods to feed this family of four for the next two weeks.

We pick up the kids from Jake's sitter, and by the time we get back to the house, Lee is home from work. As we unpack the groceries and put them away, we talk about the family's food choices at both supermarkets and restaurants. When they want a dinner out as a family, they go to El Chico for Mexican food, Smokey Joe's Barbeque, or Larry's Pizza. When Jake and Lee are by themselves and in a hurry, Lee goes to Sonic and Popeye's; Jake likes McDonald's, Burger King, and Arby's, where she favors the turkey sandwich.

After the kids are put to bed, we drink beer and talk. The main thing on their minds is the time consumed by Lee's job and the needs of two small children. Before she became a mother, Jake had a busy job as a lobbyist and administrator for an association of insurance and financial advisors. Lee used to play guitar in a local rock and roll band. They no longer have time for the boating, skiing, and camping that they once enjoyed. But there's no tone of complaining or nostalgia for more carefree days; now with a toddler and an infant they have new joys and new responsibilities. It's as simple as that.

Eventually we get around to talking about what drives their food choices.

"Price and convenience are way up there, especially now with the kids," Jake says. She goes on to explain how pregnancy made her appetites shift, how she hated eggs, had little appetite for red meat, but craved cookies. Now she's starting to eat more meat again, except for sausage. "I have sort of a disdain for pig meat," she says. Lee reminds her of bacon, which she admits to enjoying. We ask them what they know about the origins of these products and the controversies about some of the modern ways of raising cattle, pigs, and chickens.

Lee knows about the chicken farms. "They're just big, long shacks packed full of chickens. You know that just from driving around the state." Arkansas is the home base of Tyson Foods, the world's largest producer of meat chickens. Neither of them knows much about pig farms or cattle feedlots. "We don't hear much about that around here," Lee says. "Most of the cattle around here are free-range, as far as I know."

For Jake, the controversy over veal calves sticks out in her mind. "That's the first one that came up when I was growing up. Veal was definitely out, without question. I mean, it was so well covered in the media, how the calves could barely move. Eating it just didn't seem worth it for the cost to the animals . . . and the horror." She admits that she, too, is not very aware of any controversies over pig farming methods. "But the chickens concern me, because I'm well aware of that, living here in Arkansas. But there's the rub, you see. We're told by dieticians to choose chicken over red meat, for health reasons."

Jake stops for a moment, obviously thinking about something related. "To be perfectly honest about it, I do think there's a hierarchy of animals. I believe I would favor mammals over birds. I think I probably feel sorrier for a cow than I would for a chicken."

"Honestly, I don't think about it that much," Lee adds. "I guess I'm pretty absorbed in my life and my family most of the time and I don't think very much about the welfare of the meat I'm eating." Lee grew up near Little Rock, and meat was always the center of the family meals. "It was either fried chicken, mashed potatoes, fried okra, or it was sweet-and-sour meatballs or rump roast, pork loins. There was always a side of vegetables, but it was the meat that was the center of any meal." For school lunches there was usually hamburger or pizza.

Jake's formative years were spent in Florida and Washington, D.C. Her mother cooked a lot from scratch, not liking pre-packaged food. There was usually some kind of meat with vegetables and potatoes on the side, but Jake's mother also made spaghetti and a lot of Chinese stir-frys, usually with chicken, beef, or shrimp. Though Jake and Lee have started to eat more vegetables than they used to, Lee doesn't anticipate any significant change in his consumption of meat: "My own philosophy is that we evolved to become omnivores, which was one of our steps in survival and in becoming what we are today. Being opportunists, we could survive the longest winters or the desert or whatever, because we ate meat. We would eat anything. It just seems like a natural order to me."

"Well, I have more qualms about it." Jake says. "There's a feeling in me that says, OK, yeah, we're adapted to eat meat, but if we don't have to, then why do it? If it was a matter of necessity, that would be one thing. Like if we're stuck in a cave and starving to death, I'm going to cut off your leg and chow down, you know?"

"Don't go hiking with her," Lee laughs.

We talk about the demands of marriage and children and how other considerations affect their food choices. Would they choose differently if they had more information about how their food was produced? Probably not—the alternatives are inconvenient and cost more. "Laziness is part of it, too," Jake says. "There's one store here where you can be assured that everything you buy is organically grown and all the meats are free-range. Everything is politically correct for the ethical meat eater, the careful carnivore. But it's about a twenty-five-minute drive from here . . . in nasty traffic. And all of the meat there is two or three times more expensive than what I get at Wal-Mart, which is only about five minutes away." Then she pauses a moment before saying: "Isn't it a sad thing when our morals become so disposable?"

Later, driving on from Mabelvale, we ponder that line. It's easy to understand why Jake and Lee make the food choices they do. They are, as Lee said, absorbed in their family and, in his case, his work, too. Making different choices would take time and add to their food bills. It's reasonable for a couple in their situation to recoil from the prospect of paying substantially more for their food, especially when buying organically grown vegetables and free-range meats would take more time as well. Is organic food really better for you, or for the environment? It's not easy to be sure. Nothing in the television they watch or the newspapers they read suggests that there is anything unethical about the choices they are making. Doesn't all of America shop at Wal-Mart? How can it be wrong to do as everyone else does?

# 2

# THE HIDDEN COSTS OF CHEAP CHICKEN

Among the items that Jake bought on her grocery shopping trip is an icon of modern American food production: chicken. Americans eat a phenomenal amount of chicken, more than any other meat. Britons, too, now eat more chicken than any other meat, and chicken accounts for about one third of all meat production in the U.K.[1] Those of us over 50 can still remember when chicken was a treat for special occasions because it was more expensive than beef. Today chicken is the cheapest meat, and its consumption has doubled since 1970.[2] Advocates of factory farming boast that their techniques have brought chicken within the reach of working families.

The chicken breasts Jake bought were produced by Tyson Foods, a corporation that proudly calls itself "the largest provider of protein products on the planet," as well as "the world leader in producing and marketing beef, pork, and chicken."[3] Tyson now produces more than 2 billion chickens a year, and if you are shopping in a typical American supermarket, close to a quarter of the chicken you see on the shelves will have been produced by Tyson.[4] Although the corporation contracts out the actual growing of its chickens to "independent" growers who own their own land and sheds, Tyson controls every aspect of production. It hatches the chicks, delivers them to the growers, tells the growers exactly how to raise them, buys back the grown chickens, and then slaughters and processes them.

Virtually all the chicken sold in America—more than 99 percent, according to Bill Roenigk, vice-president of the National Chicken Council—comes from factory-farm production similar to that used by Tyson Foods.[5] The ethical issues raised by its production of chicken therefore exemplify issues raised by modern intensive chicken production in general. We can divide these issues into three categories, according to whether they most immediately impact the chickens, the environment, or humans.

## AN ETHICAL WAY OF TREATING CHICKENS?

To call someone a "birdbrain" is to suggest exceptionally stupidity. But chickens can recognize up to 90 other individual chickens and know whether each one of those birds is higher or lower in the pecking order than they are themselves. Researchers have shown that if chickens get a small amount of food when they immediately peck at a colored button, but a larger amount if they wait 22 seconds, they can learn to wait before pecking.[6] Moreover, after thousands of generations of domestic breeding, chickens still retain the ability to give and to understand distinct alarm calls, depending on whether there is a threat from above, like a hawk, or from the ground, like a raccoon. When scientists play back a recording of an "aerial" alarm call, chickens respond differently than when they hear a recording of a "ground" alarm call.[7] Australian researcher Lesley Rogers, professor of Neuroscience and Animal Behavior at the University of New England, in Armidale, and a leading expert on the behavior of the chicken, has said that "[I]t is now clear that birds have cognitive capacities equivalent to those of mammals, even primates."[8]

Interesting as these studies are, the point of real ethical significance is not how clever chickens are, but whether they can suffer—and of that there can be no serious doubt. Chickens have nervous systems similar to ours, and when we do things to them that are likely to hurt a sensitive creature, they show behavioral and physiological responses that are like ours. When stressed or bored, chickens show what scientists call "stereotypical behavior," or repeated futile movements, like caged animals who pace back and forth. When they have become acquainted with two different habitats and find one preferable to the other, they will work hard to get to the living quarters they prefer. Lame chickens will choose food to which painkillers have been added; the drug evidently relieves the pain they feel and allows them to be more active.[9]

Most people readily agree that we should avoid inflicting unnecessary suffering on animals. Summarizing the recent research on the mental lives of chickens and other farmed animals, Christine Nicol, professor of animal welfare at Bristol University, in England, has said: "Our challenge is to teach others that every animal we intend to eat or use is a complex individual, and to adjust our farming culture accordingly."[10] We are about to see how far that farming culture would have to change to achieve this.

Almost all the chickens sold in supermarkets—known in the industry as "broilers"—are raised in very large sheds. A typical shed measures 490 feet long by 45 feet wide and will hold 30,000 or more chickens. The National Chicken Council, the trade association for the U.S. chicken industry, issues Animal Welfare Guidelines that indicate a stocking density of 96 square inches for a bird of average market weight[11]—that's about the size of a standard sheet of American 8.5 inch × 11 inch typing paper. In the U.K., stocking densities can be as high as 19 birds per square metre of floor space. This gives each chicken less floor space than the size of a piece of A4 paper to live in.[12] When the chicks are small, they are not crowded, but as they near market weight, they cover the floor completely—at first glance, it seems as if the shed is carpeted in white. They are unable to move without pushing through other birds, unable to stretch their wings at will, or to get away from more dominant, aggressive birds. The crowding causes stress, because in a more natural situation, chickens will establish a "pecking order" and make their own space accordingly.

If the producers gave the chickens more space they would gain more weight and be less likely to die, but it isn't the productivity of each bird—let alone the bird's welfare—that determines how they are kept. As one industry manual explains: "Limiting the floor space gives poorer results on a per bird basis, yet the question has always been and continues to be: What is the least amount of floor space necessary per bird to produce the greatest return on investment."[13]

In Britain, a judge ruled in 1997 that crowding chickens like this is cruel. The case arose when McDonald's claimed that two British environmental activists, Helen Steel and David Morris, had libeled the company in a leaflet that, among other things, said that McDonald's was responsible for cruelty. Steel and Morris had no money to pay lawyers to defend themselves against the corporate giant so they ran the case themselves, calling experts to give evidence in support of their claims. The "McLibel" case turned into the longest trial in English legal history.

## ENTER THE CHICKEN SHED

*(Warning: May Be Disturbing to Some Readers)*

Enter a typical chicken shed and you will experience a burning feeling in your eyes and your lungs. That's the ammonia-it comes from the birds' droppings, which are simply allowed to pile up on the floor without being cleaned out, not merely during the growing period of each flock, but typically for an entire year, and sometimes for several years.[14] High ammonia levels give the birds chronic respiratory disease, sores on their feet and hocks, and breast blisters. It makes their eyes water, and when it is really bad, many birds go blind.[15] As the birds, bred for extremely rapid growth, get heavier, it hurts them to keep standing up, so they spend much of their time sitting on the excrement-filled litter-hence the breast blisters.

Chickens have been bred over many generations to produce the maximum amount of meat in the least amount of time. They now grow three times as fast as chickens raised in the 1950s while consuming one-third as much feed.[16] But this relentless pursuit of efficiency has come at a cost: their bone growth is outpaced by the growth of their muscles and fat. One study found that 90 percent of broilers had detectable leg problems, while 26 percent suffered chronic pain as a result of bone disease.[17] Professor John Webster of the University of Bristol's School of Veterinary Science has said: "Broilers are the only livestock that are in chronic pain for the last 20 percent of their lives. They don't move around, not because they are overstocked, but because it hurts their joints so much."[18] Sometimes vertebrae snap, causing paralysis. Paralyzed birds or birds whose legs have collapsed cannot get to food or water, and—because the growers don't bother to, or don't have time to, check on individual birds—die of thirst or starvation. Given these and other welfare problems and the vast number of animals involved—nearly 9 billion in the United States and more than 800 million in the U.K.—Webster regards industrial chicken production as, "in both magnitude and severity, the single most severe, systematic example of man's inhumanity to another sentient animal."[19]

Criticize industrial farming, and industry spokespeople are sure to respond that it is in the interests of those who raise animals to keep them healthy and happy so that they will grow well. Commercial chicken-rearing conclusively refutes this claim. Birds who die prematurely may cost the grower money, but it is the total productivity of the shed that matters. G. Tom Tabler, who manages the Applied Broiler Research Unit at the University of Arkansas, and A. M. Mendenhall of the Department of Poultry Science at the same university, have posed the question: "Is it more profitable to grow the biggest bird and have increased mortality due to heart attacks, ascites (another illness caused by fast growth), and leg problems, or should birds be grown slower so that birds are smaller, but have fewer heart, lung, and skeletal problems?" Once such a question is asked, as the researchers themselves point out, it takes only "simple calculations" to draw the conclusion that, depending on the various costs, often "it is better to get the weight and ignore the mortality."[20]

Breeding chickens for rapid growth creates a different problem for the breeder birds, the parents of the chickens people eat. The parents have the same genetic characteristics as their offspring—including huge appetites. But the breeder birds must live to maturity and keep on breeding as long as possible. If they were given as much food as their appetites demand, they would grow grotesquely fat and might die before they became sexually mature. If they survived at all, they would be unable to breed. So breeder operators ration the breeder birds to 60 to 80 percent less than their appetites would lead them to eat if they could.[21] The National Chicken Council's Animal Welfare Guidelines refer to "off-feed days;" that is, days on which the hungry birds get no food at all. This is liable to make them drink "excessive" amounts of water, so the water, too, can be restricted on those days. They compulsively peck the ground, even when there is nothing there, either to relieve the stress, or in the vain hope of finding something to eat. As Mr. Justice Bell, who examined this practice in the McLibel case, said: "My conclusion is that the practice of rearing breeders for appetite, that is to feel especially hungry, and then restricting their feed with the effect of keeping them hungry, is cruel. It is a well-planned device for profit at the expense of suffering of the birds."

The fast-growing offspring of these breeding birds live for only six weeks. At that age they are caught, put into crates, and trucked to slaughter. A *Washington Post* journalist observed the catchers at work: "They grab birds by their legs, thrusting them like sacks of laundry into the cages, sometimes applying a shove." To do their job more quickly, the catchers pick up only one leg of each bird, so that they can hold four or five chickens in each hand. (The National Chicken Council's Animal Welfare Guidelines, eager to avoid curtailing any practice that may be economically advantageous, says "The maximum number of birds per hand is five.") Dangling from one leg, the frightened birds flap and writhe and often suffer dislocated and broken hips, broken wings, and internal bleeding.[22]

Crammed into cages, the birds then travel to the slaughterhouse, a journey that can take several hours. When their turn to be removed from the crates finally comes, their feet are snapped into metal shackles hanging from a conveyor belt that moves towards the killing room. Speed is the essence, because the slaughterhouse is paid

*(continued on page 26)*

by the number of pounds of chicken that comes out the end. Today a killing line typically moves at 90 birds a minute, and speeds can go as high as 120 birds a minute, or 7,200 an hour. Even the lower rate is twice as fast as the lines moved twenty years ago. At such speeds, even if the handlers wanted to handle the birds gently and with care, they just couldn't.

In the United States, in contrast to other developed nations, the law does not require that chickens (or ducks, or turkeys) be rendered unconscious before they are slaughtered. As the birds move down the killing line, still upside down, their heads are dipped into an electrified water bath, which in the industry is called "the stunner." But this is a misnomer. Dr Mohan Raj, a researcher in the Department of Clinical Veterinary Science at the University of Bristol, in England, has recorded the brain activity of chickens after various forms of stunning and reported his results in such publications as *World's Poultry Science Journal.* We asked him: "Can the American consumer be confident that broilers he or she buys in a supermarket have been properly stunned so that they are unconscious when they have their throats cut?" His answer was clear: "No. The majority of broilers are likely to be conscious and suffer pain and distress at slaughter under the existing water bath electrical stunning systems." He went on to explain that the type of electrical current used in the stunning procedure was not adequate to make the birds immediately unconscious. Using a current that would produce immediate loss of consciousness, however, would risk damage to the quality of the meat. Since there is no legal requirement for stunning, the industry won't take that risk. Instead, the inadequate current that is used evidently paralyzes the birds without rendering them unconscious. From the point of view of the slaughterhouse operator, inducing paralysis is as good as inducing unconsciousness, for it stops the birds from thrashing about and makes it easier to cut their throats.

Because of the fast line speed, even the throat-cutting that follows the electrified water bath misses some birds, and they then go alive and conscious into the next stage of the process, a tank of scalding water. It is difficult to get figures on how many birds are, in effect, boiled alive, but documents obtained under the Freedom of Information act indicate that in the United States alone, it could be as many as three million a year.[23] At that rate, 11 chickens would have been scalded to death in the time it takes you to read this page. But the real figure might be much higher. An undercover videotape made at a Tyson slaughterhouse at Heflin, Alabama, shows dozens of birds who have been mutilated by throat-cutting machines that were not working properly. Workers rip the heads off live chickens that have been missed by the cutting blade. Conscious birds go into the scalding tank. A plant worker is recorded as saying that it is acceptable for 40 birds per shift to be missed by the backup killer and scalded alive.[24]

If you found the last few paragraphs unpleasant reading, Virgil Butler, who spent years working for Tyson Foods in the killing room of a slaughterhouse in Grannis, Arkansas, killing 80,000 chickens a night, mostly for Kentucky Fried Chicken, says that what we have described "doesn't even come close to the horrors I have seen." On an average night, he says, about one in every three of the chickens were alive when they went into the scalding tank. The missed birds are, according to Butler, "scalded alive." They "flop, scream, kick, and their eyeballs pop out of their heads." Often they come out "with broken bones and disfigured and missing body parts because they've struggled so much in the tank."[25] When there were mechanical failures, the supervisor would refuse to stop the line, even though he knew that chickens were going into the scalding tank alive or were having their legs broken by malfunctioning equipment.

In January 2003, Butler made a public statement describing workers pulling chickens apart, stomping on them, beating them, running over them on purpose with a fork-lift truck, and even blowing them up with dry ice "bombs." Tyson dismissed the statement as the "outrageous" inventions of a disgruntled worker who had lost his job. It's true that Butler has a conviction for burglary and has had other problems with the law. But eighteen months after Butler made these supposedly "outrageous" claims, a videotape secretly filmed at another KFC-supplying slaughterhouse, in Moorefield, West Virginia, made his claims a lot more credible. The slaughterhouse, operated by Pilgrim's Pride, the second largest chicken producer in the nation, had won KFC's "Supplier of the Year" Award. The tape, taken by an undercover investigator working for People for the Ethical Treatment of Animals, showed slaughterhouse workers behaving in ways quite similar to those described by Butler: slamming live chickens into walls, jumping up and down on them, and drop-kicking them as if they were footballs. The undercover investigator said that, beyond what he had been able to catch on camera, he had witnessed "hundreds" of acts of cruelty. Workers had ripped off a bird's head to write graffiti in blood, plucked feathers off live chickens to "make it snow," suffocated a chicken by tying a latex glove over its head, and squeezed birds like water balloons to spray feces over other birds. Evidently, their work had desensitized them to animal suffering.

The only significant difference between the behavior of the workers at Moorefield and that described by Butler at Grannis was that the behavior at Moorefield was caught on tape. Unable to dismiss the evidence of cruelty, Pilgrim's Pride said that it was "appalled."[26] But neither Pilgrim's Pride nor Tyson Foods, the two largest suppliers of chicken in America, have done anything to address the root cause of the problem: unskilled, low-paid workers doing dirty, bloody work, often in stifling heat, under constant pressure to keep the killing lines moving no matter what so that they can slaughter up to 90,000 animals every shift.

After hearing many experts testify, the judge, Rodger Bell, ruled that, although some other claims Steel and Morris had made were false, the charge of cruelty was true: "Broiler chickens which are used to produce meat for McDonald's . . . spend the last few days of their lives with very little room to move," he said. "The severe restriction of movement of those last few days is cruel and McDonald's are culpably responsible for that cruel practice."[27]

## A DAY IN THE LIFE OF A TURKEY INSEMINATOR

The turkey meat in the sandwiches Jake buys at Arby's would come from factory-farmed turkeys, reared in much the same way as chickens. The main difference is that because turkeys have been bred to have such an oversized breast, they cannot mate naturally. A few years ago we learned that the Butterball Turkey company, a division of the agribusiness giant ConAgra, needed workers for its artificial insemination crews in Carthage, Missouri. Our curiosity piqued, we decided to see for ourselves what this work really involved. The only qualification for the job seemed to be the ability to pass a drug test, so we were hired.

We spent some time on both sides of the job: collecting the semen and getting it into the hen. The semen comes from the "tom house," where the males are housed. Our job was to catch a tom by the legs, hold him upside down, lift him by the legs and one wing, and set him up on the bench on his chest/neck, with the vent sticking up facing the worker who actually collected the semen. He squeezed the tom's vent until it opened up and the white semen oozed forth. Using a vacuum pump, he sucked it into a syringe. It looked like half-and-half cream, white and thick. We did this over and over, bird by bird, until the syringe was filled to capacity with semen and a sterile extender. The full syringe was then taken over to the hen house.

In the hen house, our job was to "break" the hens. You grab a hen by the legs, trying to cross both "ankles" in order to hold her feet and legs with one hand. The hens weigh 20 to 30 pounds and are terrified, beating their wings and struggling in panic. They go through this every week for more than a year, and they don't like it. Once you have grabbed her with one hand, you flop her down chest first on the edge of the pit with the tail end sticking up. You put your free hand over the vent and tail and pull the rump and tail feathers upward. At the same time, you pull the hand holding the feet downward, thus "breaking" the hen so that her rear is straight up and her vent open. The inseminator sticks his thumb right under the vent and pushes, which opens it further until the end of the oviduct is exposed. Into this, he inserts a straw of semen connected to the end of a tube from an air compressor and pulls a trigger, releasing a shot of compressed air that blows the semen solution from the straw and into the hen's oviduct. Then you let go of the hen and she flops away.

Routinely, methodically, the breakers and the inseminator did this over and over, bird by bird, 600 hens per hour, or ten a minute. Each breaker "breaks" five hens a minute, or one hen every 12 seconds. At this speed, the handling of birds has to be fast and rough. It was the hardest, fastest, dirtiest, most disgusting, worst-paid work we have ever done. For ten hours we grabbed and wrestled birds, jerking them upside down, facing their pushed-open assholes, dodging their spurting shit, while breathing air filled with dust and feathers stirred up by panicked birds. Through all that, we received a torrent of verbal abuse from the foreman and others on the crew. We lasted one day.

## THE COST TO THE ENVIRONMENT

The Delmarva Peninsula, so called because parts of it belong to Delaware, Maryland, and Virginia, has great natural gifts: green rolling countryside, estuaries, beaches, and two great bodies of water, the Atlantic Ocean on the east and the vast Chesapeake Bay on the west. On the surface, some parts of Chesapeake Bay and its surrounding countryside seem to be among the few remaining nearly natural areas on the East Coast. Underneath, however, the Bay is in serious trouble. When Captain John Smith entered Chesapeake Bay in 1608, it was such a thriving natural environment that he joked you could catch fish with a frying pan. Until well into the twentieth century, the Bay was carpeted with beds of clams and oysters—a huge living filter that kept the water clean. Now the few remaining oysters can't do that.[28] Over-harvesting and the growth of human population in the region, with the pollution it brings, are partial reasons for the bay's problems; but recently attention has been turning to the chicken industry on the peninsula itself.

More than 600 million chickens a year are raised on the Delmarva Peninsula. Those chickens produce more manure than a city of four million people, and instead of getting processed like human waste, chicken manure is spread on fields. But the Delmarva cannot absorb that much nitrogen and phosphorus. Sussex County, Delaware, produces 232 million chickens annually—more than any other county in the nation—but a University of Delaware study found that the county only has enough land to cope with the manure from 64 million chickens. Up to half of the nutrients in the excess manure washes off into the rivers and streams, or gets into the groundwater. A third of the shallow wells in the Delmarva Peninsula, including those going into the underground aquifer used for drinking water, have nitrate levels above the federal safe drinking water standards, according to the U.S. Geological Survey. In the rivers and the bay, these nutrients stimulate too much algae growth. The algae decomposes, sucking oxygen out of the water, and fish and other forms of water life die. The bay now has "dead zones" that cannot support fish, crabs, oysters, or other species of ecological significance. In July 2003, a dead zone stretched for 100 miles down the central portion of the bay.[29]

In western Kentucky, the masthead of *The Messenger,* the local newspaper of Madisonville, carries the slogan "The Best Town on Earth." But if you had been in the audience of a hearing at the Madisonville Technology Center on the evening of June 29, 2000, you would have had to wonder about that. The Natural Resources and Environmental Protection Cabinet of the Kentucky Department of Environmental Protection was listening to public comment on a proposed regulation for Concentrated Animal Feeding Operations, also known as factory farms. A long procession of citizens came up and made their views known. Here is a selection:[30]

> Since Tyson took over the operation of the growing houses, there is a very offensive odor that at times has taken my breath. There has been a massive invasion of flies. It is hard to perform necessary maintenance on our property.[31]

> Uncovered hills of chicken waste attract hundreds of thousands of flies and mice... People, including school children, can not enjoy a fresh morning's air and can't inhale without gagging or coughing due to the smell.[32]

> My family lives next to chicken houses. We caught 80 mice in two days in our home. The smell is nauseating . . . My son and I got stomach cramps, diarrhea, nausea, and we had a sore on our mouths that would not go away. We went to the doctor and my son had parasites in his intestines. Where are the children's rights? Should families have to sacrifice a safe and healthy environment for the economic benefit of others?[33]

After the hearings, local western Kentucky residents, supported by the Sierra Club, sued Tyson Foods for failing to report releases of ammonia from four of its chicken factories as it was required by law to do. Tyson claimed that because the factories were owned by growers who only contracted to sell their chickens to Tyson, it was not responsible for reporting the pollution. In 2003, a federal court rejected that argument, holding that since Tyson controlled how the chickens were raised, what they were fed, and what medications they were given, and gained most of the profit from raising them, Tyson was also responsible for the pollution.[34] Tyson finally settled with the residents in 2005, agreeing to spend $500,000 to study and report on emissions and mitigate ammonia emissions. Tyson also agreed to pay the legal costs of the residents and to plant trees to screen the chicken factories and reduce odors. Sierra Club attorney Barclay Rogers hailed the outcome as a "landmark decision" that has established the responsibility of factory farms to "clean up their act and stop putting communities at risk."[35]

The Delmarva Peninsula and western Kentucky are examples of a nationwide problem. In Warren County, in northern New Jersey, Michael Patrisko, who lives near an egg factory farm, told a local newspaper that the flies around his neighborhood are so bad, "You literally can look at a house and think it's a different color."[36] Buckeye Egg Farm in Ohio was fined $366,000 for failing to handle its manure properly. Nearby residents had complained for years about rats, flies, foul odors, and polluted streams from the 14-million-hen complex.[37] At the same time, Oklahoma Attorney General Drew Edmondson was threatening to sue Arkansas poultry producers, including Tyson Foods, saying that waste from the companies' operations is destroying Oklahoma lakes and streams, especially in the northeast corner of the state.[38]

Tyson Foods, which the Sierra Club listed as one of the Ten Least Wanted Animal Factory Operators in 2002, has a long history of convictions for pollution.[39] After the incidents the Sierra Club listed in its 2002 report, Tyson was again in the news in 2003, when it admitted that it had repeatedly discharged untreated wastewater from its poultry plant in Sedalia, Missouri into a tributary of the Lamine River. The plant, which covers a thousand acres and processes about a million chickens a week, discharges hundreds of thousands of gallons of wastewater every day. State and federal prosecutors alleged that over the previous decade Tyson had repeatedly ignored civil fines, state orders, and other violation notices about its wastewater discharges. Tyson acknowledged that employees at the plant knew about the discharges and agreed to pay a total of $7.5 million in fines.[40]

Tyson produces chicken cheaply because it passes many costs on to others. Some of the cost is paid by people who can't enjoy being outside in their yard because of the flies and have to keep their windows shut because of the stench. Some is paid by kids who can't swim in the local streams. Some is paid by those who have to buy bottled water because their drinking water is polluted. Some is paid by people who want to be able to enjoy a natural environment with all its beauty and rich biological diversity. These costs are, in the terms used by economists, "externalities" because the people who pay them are external to the transaction between the producer and the purchaser.

Consumers may choose to buy Tyson chicken, but those who bear the other, external costs of intensive chicken production do not choose to incur them. Short of moving house—which has its own substantial costs—there is often little they can do about it. Economists—even those who are loudest in extolling the virtues of the free market—agree that the existence of such externalities is a sign of market failure. In theory, to eliminate this market failure, Tyson should fully compensate everyone adversely affected by its pollution. Then its chicken would no longer be so cheap.

## THE COST TO WORKERS

Jobs at Tyson Foods are so poorly paid and unpleasant that job turnover in some plants has been reported to be higher than 100 percent annually, meaning that the average employee lasts less than a year—although Tyson refuses to make the figure public.[41] Some of the jobs are also dangerous. In 1999 Tyson Foods was named one of the Ten Worst Corporations of the Year by *Corporate Crime Reporter*. That year, seven Tyson workers died in industrial accidents. One of them was a 15-year-old boy working as a chicken catcher in Arkansas. In the same year, another 15-year-old was seriously injured in an accident at a Tyson plant in Missouri. As a result of the accidents, two 14-year-olds and another 15-year-old were discovered working at the Missouri plant. Tyson was fined by the Department of Labor for violating the child labor provisions of the Fair Labor Standards Act.[42] The corporation was also fined by the Occupational Health and Safety Administration for violating health and safety laws in several states.[43]

Tyson has a record of seeking to lower wages and cut health benefits for its workers, even while the corporation has been experiencing unprecedented growth and making billions of dollars in profits. When Tyson took over IBP, a major beef producer, it found that it had acquired a workforce receiving better wages and benefits than its own workers—and it set out to change that. In February 2003, Tyson offered workers at a former IBP plant in Jefferson, Wisconsin, a new contract that included pay cuts, no pensions for new workers and frozen pensions for existing workers, cuts in vacation time, and higher health insurance co-payments for an inferior health care package. The plant had been profitable when run by IBP and was continuing to make a profit for Tyson—but not as big a profit as Tyson thought it could make. A company manager said that the workers' pay of $25,000 to $30,000 a year, plus benefits, made the plant an "outlier" and put its workers "in a luxurious position from our perspective."[44] A long, bitter strike ensued, but after Tyson brought in new workers willing to cross the picket lines, strikers eventually had to accept virtually the same deal that they had been offered before the strike began.

The contractors who are responsible for rearing the chickens, known in the industry as "growers," may seem to be more independent than employees, but once they have signed on, they have little choice but to take the terms Tyson offers them. Growers have to invest their own money in the sheds and equipment and often go heavily into debt to do so. They then become dependent on constantly renewing their contract with a corporation like Tyson to get their money back—for without a new contract, no more chicks will arrive in their sheds. Since the sheds are useless for anything but growing chickens, the growers can lose not only their investment, but their land as well. Usually only one corporation operates within a 25-mile radius, and even if two corporations are operating in the same area, there is often an unwritten rule that one company will not pick up a grower who has worked for another company. So if a grower does not like the contract that Tyson offers, there is nowhere else to go. Their independence gone, the growers are, as one of them put it, "serfs at the mercy of the companies that make a fortune on their backs."[45]

Corporations often defend their low wages by saying that if people don't like the pay they are offering, they don't have to take it. Employees in poultry slaughterhouses can, they say, seek work elsewhere, and growers are free to grow chickens and sell them themselves, if they can—or not get into chicken growing in the first place. In a free market, that's how things work, and the consumers benefit in terms of lower prices. However, the job options available to many low-skilled chicken growers may be limited. In any case, this argument doesn't excuse the mistreatment of chickens who, unlike workers, have no choice at all. In the end, consumers, too, are free to choose. If they don't like the way a corporation treats its workers and contractors, or the environment, or the animals it uses, they can take their money elsewhere.

## THE GREATEST COST OF ALL?

In 2005, the world began to face the serious possibility that the cheap chicken produced by factory farming could be far more costly to all of us than even the most radical animal rights advocates had ever dreamt it might be. Scientists began to warn leaders to prepare against the possibility of an epidemic of avian influenza—popularly known as bird flu—that could spread to human beings and take tens, or even hundreds, of millions of lives. Supporters of factory farming have used the threat of bird flu to make a case against having chickens outdoors, claiming that the virus can be spread by migrating birds to free-range flocks.[46] But the real danger, as scientists now recognize, is intensive chicken production.

In October 2005, a United Nations task force identified as one of the root causes of the bird flu epidemic, "farming methods which crowd huge numbers of animals into small spaces." [47] Other experts agree, among them University of Ottawa virologist Earl Brown, who said after a Canadian outbreak of avian influenza, " . . . high intensity chicken rearing is a perfect environment for generating virulent avian flu virus."[48] Although transmission through wild birds to chickens kept outdoors is a possibility, as Dr Brown has pointed out, viruses found in wild birds are generally not very dangerous. It is when they get into a high-density poultry operation that they mutate into something much more virulent. Traditionally reared birds, moreover, are likely to have more resistance than the stressed, genetically similar birds kept in intensive confinement systems. And in any case, even if there were no chickens kept outside, factory farms are not biologically secure. They are frequently infested with mice, rats, and small birds who can bring in diseases.

As of this writing, the number of human beings who have died from the current strain of avian influenza, known as H5N1, is relatively small, and it appears that they have all been in contact with infected birds. But if the virus mutates into some form that is transmissible from human to human, as experts say it might, the number of deaths could outstrip the estimated 20 million victims of the Spanish flu epidemic of 1918. Governments are, rightly, taking action to prepare for this threat. In 2005, the United States Senate approved the spending of $8 billion to stockpile vaccines and other drugs to help prevent a possible bird flu epidemic. Other governments have already spent tens of millions for that and other preventive measures.

Such government spending is really a kind of subsidy to the poultry industry and, like most subsidies, it is bad economics. Factory farming spread because it seemed to be cheaper than more traditional forms of farming. We have seen that it was cheaper to the consumer, but only because it was passing some of its costs on to others—for example, to people who lived downstream or downwind from the factory farms and could no longer enjoy clean water and air and to workers who were injured by unsafe conditions. Now we can see that this was only the small stuff. Factory farming is passing far bigger costs—and risks—on to all of us. If chicken were taxed to raise enough revenue to pay for the precautions that governments now have to take against avian influenza, again we might find that factory farm chicken isn't really so cheap after all.

## A CLEAR-CUT CONCLUSION

Gandhi remarked that the greatness of a nation and its moral progress can be judged by the way it treats its animals. If we apply that standard to industrial chicken producers, they don't come off well. Chicken producers

# 3

# BEHIND THE LABEL: "ANIMAL CARE CERTIFIED" EGGS

The carton of Country Creek eggs that Jake Hillard picked up at Wal-Mart carried the name Moark Productions, one of America's largest producers of eggs. It also bore a red seal saying Animal Care Certified. We asked Jake if the seal signified anything to her. "Well, it seemed to imply that they followed some standard of humane animal care," she said. "I get the general impression that the chickens are cared for better than by some companies, but I don't know by how much."

Jake's vagueness about the Animal Care Certified seal wasn't surprising. Most Americans know little about how their eggs are produced. They don't know that American egg-producers typically keep their hens in bare wire cages, often crammed eight or nine hens to a cage so small that they never have room to stretch even one wing, let along both. The space allocated per hen, in fact, is even less than broiler chickens get, ranging from 48 to 72 square inches. Even the higher of these figures is less than the size of a standard American sheet of typing paper. In such crowded conditions, stressed hens tend to peck each other—and the sharp beak of a hen can be a lethal weapon when used relentlessly against weaker birds unable to escape. To prevent this, producers routinely sear off the ends of the hens' sensitive beaks with a hot blade—without an anaesthetic.[1] In the European Union, regulations require that hens have 550 square centimetres of space, slightly more than is standard in the U.S., but still less than a sheet of A4 paper. Debeaking of hens without anaesthetic is standard procedure for British egg producers.

As for the cages themselves, they are in long rows, sometimes stacked three and four tiers high. That way, in a single building, tens of thousands of hens can be fed, watered, and have their eggs collected by machines. Artificial lighting is used to mimic the longest days of summer, to induce the hens to lay the maximum number of eggs all year round. A year of this leaves the hens debilitated, and they start to lay fewer eggs. Many American producers then cut off their food and starve them for as long as two weeks until they go into molt, which means they lose their feathers and cease to lay eggs. Some die during this period, and the survivors lose about 30 percent of their body weight. They are then fed again, and their laying resumes for a few more months before they are killed.

Although animal advocates have been describing these conditions since the 1970s, until recently the American media have ignored them. That is changing, and much of the credit for that change must go to Paul Shapiro and Miyun Park, two young activists who at the time ran an organization called Compassion Over Killing. Paul learned about factory farms when he was 14 years old, and he started COK as a club at his high school. The club outgrew high school and attracted volunteers, among them Park, who became president a year later. The two led fur protests, sit-ins, and plenty of in-your-face street activism.

Troubled by the knowledge that within a 100-mile radius of where they lived millions of hens were suffering in cages, unseen by the people who bought the eggs the hens laid, Shapiro and Park tried a different tactic. In 2001 they began driving around rural Maryland locating egg factory farms by day and entering them with video cameras by night. Their videos show dead hens rotting in cages, hens with necks and feet caught in the wire mesh, and hens who had fallen into the manure pit beneath the batteries of cages.[2] They also show COK members gently holding sick and injured birds and taking them away to get veterinary care. This was powerful stuff, and it won Park and Shapiro the attention of writers at *The Washington Post*. The paper's exposé opened the door for a string of favorable stories about COK's open rescues in *The New York Times* and other national media.

Throughout the media brouhaha, no COK members were ever charged with trespassing or theft of birds, presumably because the egg companies did not want to acknowledge that the videos had been taken in their sheds.

There was something different about this kind of animal welfare activism, and it helped win the sympathy of the media. Shapiro explains it like this: "We were regular people who were acting in the only decent way that you could when faced with such egregious cruelty. We weren't damaging property, we weren't hiding our identities. We just simply went in there and videotaped ourselves providing aid to sick and injured animals."

Once Shapiro and Park had opened up the issue, reporters had no difficulty in finding credible experts who could attest to the conditions inside the egg factories. McDonald's has called Dr. Temple Grandin, a "preeminent animal behavior expert" and taken her advice on animal welfare issues.[3] About the egg industry, she was characteristically plain spoken:

> When I visited a large egg layer operation and saw old hens that had reached the end of their productive life, I was horrified. Egg layers bred for maximum egg production and the most efficient feed conversion were nervous wrecks that had beaten off half their feathers by constant flapping against the cage. . . . The more I learned about the egg industry the more disgusted I got. Some of the practices that had become "normal" for this industry were overt cruelty. Bad had become normal. Egg producers had become desensitized to suffering.[4]

United Egg Producers, the industry trade association representing most of the country's egg production, was concerned about the bad publicity the egg industry was getting. Its experts must also have been well aware that the entire European Union—twenty-five nations, with a much larger population of both humans and hens than the United States—was in the process of phasing out the battery cage, insisting that all hens have a place to perch, litter to scratch in, a nesting box to lay their eggs in, and about twice the space that most U.S. hens are granted. As for starving hens in order to force them to molt, that had long been illegal in the European Union. But United Egg Producers didn't recommend that its members follow Europe's example. It opted for a few minor changes and plenty of spin. Egg producers who followed a new set of voluntary guidelines would be allowed to stamp their egg cartons with a colorful seal stating that the eggs were "Animal Care Certified."

But the new guidelines were only a marginal improvement on the existing situation. They allowed each hen 67 square inches of space—by 2008. Dr Joy Mench, professor of animal science at the University of

California, Davis, and a member of UEP's own Advisory Committee, is on record as calling 80 inches a "meager" space allowance that is "barely enough for the hen to turn around and not enough for her to perform normal comfort behaviors."[5] The UEP program also permits producers to continue to sear off part of the beaks of their chickens with a hot blade, without pain relief.[6] A chicken's beak is its major organ for interacting with the ground and for picking up seeds or worms, and it is full of nerve endings. Professor Ian Duncan, who holds a chair of animal welfare at the Department of Animal and Poultry Science at the University of Guelph, Ontario, and has done decades of research on the welfare of chickens, says that "beak trimming leads to both chronic and acute pain."[7] When asked on National Public Radio what she thought about the procedure, UEP's advisor Professor Joy Mench pointed out that for chickens, their beak is their main way of exploring, touching, and feeling things. "So," the interviewer asked, "cutting off the beak is a big deal, if you're a hen?" Mench replied: "It's definitely a big deal."[8]

At the time of Jake's egg purchase, the UEP "Animal Care Certified" guidelines also permitted starving birds to make them molt. It doesn't really take an expert to say that this is going to make them suffer. In the National Public Radio interview, Mench didn't resort to scientific jargon: "The bird is starved. Yes, the bird is starved. I don't like to see hungry animals not being given food."

Finally, we shouldn't forget that the eggs that produce laying hens also produce equal numbers of male chicks. Since male chicks don't lay eggs, the egg industry doesn't want them. The broiler industry doesn't want them either, for they are not bred to gain weight rapidly, as broiler chickens are. Temple Grandin discovered what many hatcheries do with them: "They were throwing live animals in the dumpster to get rid of them. I was going, 'What? They were doing what?' Nobody would throw a live calf in a dumpster. These people forgot this is a live animal."[9] The UEP guidelines don't require producers to avoid buying from hatcheries that use this method of disposing of male chicks.

## CERTIFIED WHAT?

In 2002, when UEP announced that they would release a set of standards for animal welfare, Paul Shapiro and Miyun Park were hopeful.

"We were naïve enough," Shapiro says, "to think that they might voluntarily reform. Then we read the guidelines and saw that they would permit barren battery cages, beak searing, and forced molting through starvation." Now Shapiro and Park were even more outraged than before: "This was not just a case of animal cruelty, it had become a case of consumer fraud," Shapiro says. So they decided to go back to some of the egg farms where they had done their open rescues a year or so before. "We knew what the conditions were like back then and now here they are 'Animal Care Certified', so we thought, 'OK let's see if there's been any change'. We found the conditions were exactly the same. There is no noticeable difference between the photos of 2003 and those of 2001."

In June 2003, COK filed petitions with the Better Business Bureau objecting to UEP'S "Animal Care Certified" logo as false advertising. After examining documents submitted by UEP and COK, the Better Business Bureau ruled that the "Animal Care Certified" seal was misleading and should be discontinued. The egg trade group appealed, but the appeal board upheld the earlier ruling. More months passed, and UEP made no changes in its "Animal Care Certified" program. In August 2004, the Better Business Bureau determined that UEP was failing to comply with its ruling and formally referred the matter to the Federal Trade Commission for law enforcement action.

The pressure on the egg producers was mounting. In May 2005, UEP announced that it was recommending that egg producers switch to a molting process that does not involve starving hens and that this recommendation would, from January 1, 2006, become a requirement of a new animal care certification program. Producers should use a feed with lower protein levels instead of taking away food entirely, UEP now said.[10] Then, in September 2005, after being "encouraged" by the Federal Trade Commission to do more, the egg trade body announced it was dropping the "animal care certified" logo and replacing it with one saying "United Egg Producers Certified: Produced in compliance with UEP animal husbandry guidelines."[11] That may be literally accurate, but many consumers will still assume it means good animal welfare, when the truth is very far from that.

# 4

# MEAT AND MILK FACTORIES

The average American eats more than 200 pounds of red meat, poultry, and fish per year. That's an increase of 23 pounds over 1970, and it would be difficult for anyone to maintain that Americans in 1970 were not eating enough of these foods. In the last 35 years, the amount of beef eaten has fallen, but that has been outweighed by the near-doubling of chicken consumption. Pork comes in third, at 51 pounds per person, behind chicken and beef. More than 60 percent of the pork eaten by Americans is bought already processed, as bacon, ham, lunch meats, hot dogs, or sausage.[1] In the U.K., pigmeat products rank second, after chicken but ahead of beef, which in turn is far ahead of mutton and lamb.[2]

The Oscar Mayer bacon that Jake bought is in this category. We wanted to trace it back to the farms that raised the pigs, but that proved impossible. Oscar Mayer is now owned by Kraft Foods, the largest food and beverage company in North America and the second largest in the world (only Nestlé is bigger).[3] After numerous phone calls that involved working our way through seemingly endless menu options, we spoke to Consumer Services' Renee Zahery, who told us that "information about our procurement and processing of our product is considered proprietary in nature" and suggested we take up these questions with "a great source," Janet Riley, senior vice-president at the American Meat Institute.[4]

When we talked with Riley, she told us only that Oscar Mayer probably has contracts with suppliers such as Tyson, Smithfield, and some of the lesser-known, vertically integrated pork producing companies. So although we could not identify any of the specific farms that produce pigs for Oscar Mayer, it seems a fair assumption that their bacon comes from a cross section of today's intensive pork industry. What is that industry like?

## THE POOP ON PIGS

When Peter first wrote about factory farming in America in 1975, there were more than 660,000 pig farms producing just under 69 million pigs a year.[5] Over the next thirty years, nearly 90 percent of those pig farms vanished, so that by 2004 there were only 69,000. But these farms will produce 103 million pigs a year.[6] Across the country, the family pig farmer has been replaced by Smithfield, ConAgra, ContiGroup, and the Seaboard Corporation. Most pigs raised today come from factory farms.

The boom in mega-piggeries has caused environmental problems even more acute than those caused by intensive chicken production. An adult pig produces about four times the amount of feces of a human, so a large confinement operation with, say, fifty thousand pigs, creates half a million pounds of pig urine and excrement every day. That's as much waste as a medium-sized town—but remember that human sewage is elaborately treated before being released into the environment and factory farm waste is not.

The summer of 1995 was wetter than usual in North Carolina. During the preceding 15 years, pig production in that state had boomed, making the state the second largest pork producer in the United States. Its pigs were producing 19 million tons of waste per year—or 2.5 tons of feces and urine for every citizen in the state.[7] During that wet summer, spilled animal waste killed ten million fish in North Carolina. In one of the most dramatic incidents, an eight-acre waste pond—the industry term is "lagoon" but that word conjures up images of blue water around a coral island, not a vast outdoor cesspool—burst, releasing 25 million gallons of liquid pig excrement into the New River, killing thousands of fish and polluting the river for miles downstream. Regulations in North Carolina were tightened, but spills continue to happen from time to time across the country. Even when there is no major spill, there is often seepage from the waste pond and run-off into the creeks when the manure is sprayed onto nearby farmland.[8]

Pig factory farms are, if anything, even worse neighbors than the chicken factory farms we described earlier. Carolyn Johnsen, a reporter for the Nebraska Public Radio Network, covered the controversial growth of mega-piggeries in that state. She attended heated public meetings, divided between those who saw economic opportunities in the new industry and those angry at the contamination of their air and water and concerned about the fate of the family farm. She spoke to people like Janie Mullinex, who lives south of Imperial, Nebraska, about a mile away from 48,000 pigs confined in 24 large barns. The owner had claimed, before putting up the confinement operation, that he had new technology and it would not smell. But Mullinex claimed that was not the case. "It comes in the house—even with the windows shut, it comes in with a strong south wind," Mullinex says. "It gives my seven-year-old diarrhea if we have it all day and it makes me sick. I don't vomit, but I'm nauseous and I have a tremendous headache." The Mullinex family has new storm windows and new siding and they have insulated their house, but it hasn't stopped the smell coming in.

Johnsen also visited Mabel Bernard, who has lived and raised her family on her property near Enders, Nebraska, since 1926. Her enjoyment of her home has been spoiled by the construction of sheds holding 36,000 pigs about a mile north. When the wind comes from that direction, the stench wakes her up at night, burning her eyes and making her feel sick.[9] One Nebraska pig producer gave implicit support to those who don't want pig farms nearby when he won a 30 percent property tax reduction on his house by arguing that its value was decreased because it was located near a pig farm—his own.[10]

But big pig farms are more than a nuisance. They are also a public health risk, according to the American Public Health Association, the largest body of public health professionals in the United States. In 2003, citing a host of human diseases linked to farm animal waste and antibiotic use, the APHA passed a resolution urging government officials to adopt a moratorium on the construction of new factory farms.[11]

## A PIG'S LIFE

Pigs are affectionate, inquisitive animals. The film *Babe* was on solid scientific ground when it made its hero capable of doing everything a dog can do in the way of herding sheep. In fact, Professor Stanley Curtis thinks that the sheepdog's job would be a "pushover" for pigs he has investigated. Curtis is a hard-nosed scientist who worked for many years in the Department of Animal Sciences at the University of Illinois and received a Distinguished Service Award from the National Pork Producers Council in 2001. He conceived the idea of making it possible for pigs to tell producers what kind of conditions they prefer, and to that end, trained them to operate joystick-controlled video games. They learned quickly, and Curtis discovered that "there is much more going on in terms of thinking and observing by these pigs than we would ever have guessed."[12] The big problem, in fact, is not getting pigs to tell us what they prefer, but persuading the producers to give it to them.

To keep a dog locked up for life in a crate too narrow for her to turn around or walk more than a step or two forwards or backwards would be cruel and illegal. Yet when it comes to how pigs are kept in the U.S., here are two startling, and critical facts:

1. There is no federal law governing the welfare of farmed animals on the farm. Literally, nothing. In the U.S., federal law begins only when animals are transported or arrive at the slaughterhouse. (And even then, there is no law regarding the slaughter of chickens or other birds, who make up 95 percent of all land animals slaughtered in the U.S.) This is not because there is any constitutional barrier to covering the welfare of animals on farms, but simply because Congress has never chosen to enact any such law.
2. Most states with major animal industries have written into their anti-cruelty laws exemptions for "common farming practices." Effectively, then, cruelty is legal as long as it is done by most farmers, and you can't prosecute anyone for it.

Together, these two points mean that, as lawyer and author David Wolfson puts it, "farmed animals in such states are literally beyond the law and any common practice, no matter how horrifying, is legal."[13]

More than 90 percent of pigs raised for meat today are raised indoors in crowded pens of concrete and steel. They never get to go outside or root around in pasture and don't even have straw to bed down in.[14] The most tightly confined of all are the breeding sows. Under the factory's rigid production schedule, they are made to produce litter after litter as quickly as possible, which means that they are pregnant for most of their lives. During their pregnancies, which last about 16 weeks, most American sows are confined in "gestation crates"—steel-barred crates or stalls just a foot or so longer than their bodies, and so narrow that the sows cannot even turn around. Of the 1.8 million sows used for breeding by America's ten biggest pig producers, about 90 percent are kept in this manner, and for the industry as a whole, the figure is around 80 percent.[15]

In these conditions, apart from the brief period when they are eating, these sensitive, intelligent, and highly social animals have nothing to do all day. They cannot walk around or socialize with other sows. All they can do is stand up or lie down on the bare concrete floor. When the time comes to give birth, they are also confined in what producers call a farrowing crate. (Is it part of the gulf we draw between ourselves and other animals that leads farmers to talk of animals as "farrowing" rather than "giving birth," "feeding" rather than "eating," and "gestating" rather than "being pregnant"?) The farrowing crate keeps the sow in position, with her teats always exposed to her piglets. She is unable to roll over—and this, the defenders of the crate say, ensures that she will not roll on top of, and perhaps smother, her piglets.

In Europe, widespread public concern about the close confinement of sows led to the European Union asking its scientific veterinary advisory committee to investigate the impact of gestation crates—or sow stalls, as they are known there—on the welfare of the sows. The investigation found that sow stalls had "major disadvantages" for welfare. Pigs like to forage and explore their environment. In natural conditions they will spend up to three-quarters of their waking hours doing this. In stalls, of course, they cannot. When a sow is first put into a stall, she typically tries to escape and may push against or attack the bars. After a time, she gives up, and often becomes quite inactive and unresponsive. This, the scientific veterinary committee says, indicates clinical depression. Other sows in stalls carry out meaningless, repetitive motions, like biting the bars of the stall, chewing the air, shaking their heads from side to side, nosing around repeatedly in the empty feed trough. These pointless movements are signs of stress, similar to the endless back and forth pacing of tigers and other big cats when kept in the traditional sterile cages of old-fashioned zoos. Fortunately, many zoos have become more enlightened and no longer keep their animals in such cages. No doubt public disapproval helped persuade them to make the change. Sows in factory farms are actually worse off than the big cats in zoos used to be, because they can't even pace back and forth. But they are invisible to the public.

In addition to psychological stress, sows in crates are also less healthy than sows able to walk around. (That shouldn't be a surprise to anyone who knows that it is healthy to get some exercise.) Sows in crates frequently become lame and develop foot injuries from standing on concrete for every moment when they are not lying down. They also get more urinary tract infections.

In sum, the scientific veterinary committee concluded, "sows should preferably be kept in groups."[16] After considering this report, the European Union passed a law phasing out sow crates by the end of 2012, except for the first four weeks after mating, and requiring that sows be given straw or similar materials that they can play around with, to reduce the stress of boredom. This law will apply to all 25 countries of the European Union, which together slaughter more than twice as many pigs as the United States.[17] Even before the new law comes into effect, Britain and Sweden acted to ban sow stalls. All of the 600,000 breeding sows in Britain now have, at least, room to turn around and can interact with other pigs.

## "WAYNE BRADLEY," IOWA PIG PRODUCER

As we mentioned when discussing how agribusiness corporations refused our requests to see how they keep their animals, one Iowa pig producer was more open than all the rest. In his view, "education is the best defense against the animal rights attack on the livestock industry." He felt he had an obligation to show people around his farm, he told us, because many years ago he had became "unglued" by a television show about farming. His wife told him: 'I'm not going to listen to this. Next time somebody calls out here for an interview . . . you better talk to them. Either that or just shut up." He's been talking ever since. He talked to us several times—by telephone and in person when we toured his farm. Everything went well until we sent him what we had written and asked him to check it for any inaccuracies. At that point he suddenly asked us to not use his real name or say anything that might identify his farm and location. He had worries about "animal rights people," he said, "doing damage to things." In what follows, therefore, everything is as it happened, but we have changed the farmer's name.

The Bradley farmhouse and main buildings stand near the intersection of two county roads. Like many farmsteads in Iowa, it is sheltered from the winter winds by rows of lush cedar trees along the north and west. There is a big white house, a wide yard, silos, and an old barn. But these emblems of an older way of farming are overwhelmed by those of the new. The driveway opens up on an array of tractors, trucks, and machinery and, farther down, rows of low metal pig confinement buildings. A complex of metal grain bins, augers, and pipes towers over everything. Wayne Bradley greets us in the driveway. He is a big, hefty man, 50-something, full of energy, friendly and talkative. We walk to a small office in the corner of one of the pig buildings and sit down. Things are a bit tense at first, but grow easier as he tells us about his family and farm. He farms the land of his father and grandfather—Bradleys have been farming here since 1875. His son, Alex, farms with him and runs a herd of cattle on land of his own. It is a family farm, he says, but also a corporation because of financial advantages that incorporation offers. He farms 2,600 acres, much of it land rented from neighbors "scattered around about nine miles." The Bradleys have 500 sows, and they sell between 10,000 and 12,000 pigs a year.

As we chat, he is eager to make a few points right away. He emphatically opposes the claim that pigs in confinement are abused. "When it's thirty below zero, my hogs are laying out comfortable in a seventy-degree building—granted, it's not bedded, but they're clean and they're just laying there grunting and oblivious to the blizzard that's going on outside. That's as opposed to when we used to raise them out in open sheds and we spent the day bedding them and they'd have frozen ears and frozen tails and those types of things."

Wayne wants us to understand the economics that have driven his decisions. "We've had to specialize to a certain degree," he says. His other major concern is government regulation, primarily of waste handling. He feels that his farm is "under intense regulation and intense scrutiny all the time." He believes it is unfair because he collects manure and wastes in a concrete basin and can use the nutrients on the fields as weather permits. "Our capability of handling wastes now is so much better than before. Our chances of polluting are so much less because we inject it." Many pig producers mix the manure with water and spray it onto fields. That just leaves it on the surface where it can easily run off into creeks when it rains. Wayne has a liquid manure injector, which he pulls behind a tractor. Essentially a large tank on wheels, it pumps liquid manure down on the ground where discs cover it with about two inches of topsoil.

Wayne obviously feels caught in the middle—being squeezed between those who promote organic or pasture-raised pork and the giant corporations that now dominate pig production. "I'll defend confinement. I'm not going to defend Smithfield Foods because I think it's taking it to the extreme. They had 250,000 sows and then they went to 500,000 and now they're up to 700,000 sows. I don't think that's economically healthy . . . The packers are getting more power and control than they need."

## MAKING BACON

The Bradleys' pigs are in what's known as "total confinement"—none of them ever go outdoors. He begins the tour by taking us into one of his four farrowing rooms, where his sows give birth and then feed their piglets. These were his first confinement buildings, which he built himself in 1975. He tells us, "I was so happy when we got the hogs in here. I could get them in out of the cold." We are in a large room maybe twenty by about forty feet. It stinks inside, of course, but not as badly as some units we've been in over the years. A concrete walkway runs down the length of the building between two rows of farrowing crates containing sows and baby pigs. The crate has two parts: a taller metal framework to hold the large sow and a lower "creep" area to one side where the baby pigs sleep when they are not nursing. The sow's part is about two feet by six feet; her body nearly fills the space. She can stand up and lie down to sleep or nurse her piglets. She cannot turn around or do much else. In some crates, the "floors" are steel slats; in others, large-gauge wire mesh coated in plastic. There is no straw or other soft bedding material. Pig wastes pass through the openings and fall into a shallow pit below. A system of cables and scrapers periodically sweeps the wastes down to a pipe and they flow into a covered pit outside.

Each sow stays in her farrowing crate for about 20 days. Wayne tells us that the crate offers the piglets a safe area away from the sow when she lies down to sleep or nurse them. We look down on a sow with a litter of baby pigs all piled up like puppies and fast asleep and say something about how cute they are. "Do they look like they're abused?" he asks. No, they certainly don't, we tell him. But what about the various mutilations that we had heard are routinely carried out on pigs kept in confinement: cutting off their tails and clipping their "needle teeth" and castrating them without an anaesthetic? There are reasons for each of them, Wayne explains. The pigs' needle teeth can cut their mother's nipples and they can cut each other in fighting over nipples. "Tail docking" prevents pigs from biting and chewing on each other's tails. We press him further: isn't it only pigs in confinement who bite each other's tails? Don't they do this because they are bored, spending all their time crowded together in a sterile environment with nothing to do all day long? "I guess I would have to agree with that to a point. But we used to raise pigs out in large pens like cattle lots and we had tail-biting then too. So we've been docking tails for quite a number of years." We've seen cattle lots, and we would not be surprised if the pigs were bored there too. But we keep that thought to ourselves.

Wayne castrates his male pigs at ten days after birth. Consumer demand drives that, Wayne says. Meat from male pigs with testicles has a distinctive gamy taste called "boar taint" that consumers, apparently, don't like. If the pigs are killed at an earlier age, as happens in some other countries, this isn't a problem. But the U.S. consumer likes large cuts of meat that can only come from a more mature pig, and then the taint becomes more noticeable.

Why are these painful procedures done without any anaesthetic? Again, Wayne is disarmingly candid: "I guess I don't have a good answer for that." We ask if it is the expense involved. "Well, it would be an expense. Obviously it is going to cost money. I have no idea. I can't sit here and say, 'Well it's going to cost me a dollar a pig.' Because if it was a dollar a pig, I mean there's not a dollar a pig to throw away. If it was a nickel or a penny or something like that, there would be no reason that we couldn't. But I doubt that it would be that inexpensive." We ask Wayne if he has ever heard of anyone using a local anaesthetic for these procedures: "I never have. It's obviously a question to be asked." He hesitates before continuing: "You know, maybe farm folks are more . . .

I don't know if I'd say immune to that or not. I mean until I was 22 years old my dentist never used novocaine. I went to the dentist and I grabbed ahold of the chair and he drilled and it was over."

We're thinking that we would have made a different choice—and perhaps the pigs would too, if they could—when Wayne turns the conversation back to the sow with her piglets in the farrowing crate in front of us. "Another advantage to this versus a pasture farrowing situation is that we can do a better job of keeping an eye on the sow and the pigs. If there's a problem, you're right here. It's very easy to give her a shot if she's not feeling well."

Wayne's piglets are weaned when they are two weeks and a few days old. In more natural environments, piglets nurse from their mothers for at least nine weeks, and sometimes longer,[18] but nursing would prevent the sow becoming pregnant again during that period, thereby reducing her productivity. So the piglets are removed from their mother and she goes back to the breeding area, while they are placed together with other litters in a "nursery" building on a nearby farm. The breeding area is part of the gestation room, and that is where Wayne takes us next. At one end stand three huge, hairy boars, one to a stall. Wayne explains that they stay in these stalls about half the time, spending the other half in a resting pen where they do have room to walk around. They rotate the boars back and forth, he says, because "overuse" lowers semen quality. The boars are rough and wild-looking.

A sow will be made pregnant again as soon as she comes into estrus. Wayne uses a combination of "live mating"—a boar is allowed to mount the sow—and "AI," or artificial insemination. Wayne's pregnant sows live in group pens instead of the narrow crates that are typical of the big corporate pig factories. Each of the three pens here holds up to forty sows. Each pen has an automated self-feeder in the center; it looks like another kind of crate but with gates on each end. It holds one sow at a time. Wayne explains how it works: "The sows are all tagged with an electronic chip. When one goes in there, that machine reads the chip and it tells whether or not she's had her feed for the day. They're allowed so many pounds. They can go through there until they've eaten their daily quota." The purpose, he says, is to make sure that every sow gets to eat her ration at her own speed.

We move on past the pens of pregnant sows and down a corridor. We stop at a steel door with a small window. Wayne motions for us to take a look. It's the room where the herd manager collects semen from the boar. We ask the obvious question: "How do you collect sperm from the boar?" Wayne is all business: "We use a steel dummy." He leads us a few steps to a dusty, windowless cubby hole just off the corridor. It is about seven or eight feet square and empty except for a low steel bench with a rubber mat under it. "There's the dummy. Some of the boars will jump right on that and ride it and ejaculate. And others won't. You have to use a sow. The herdsman catches the semen in a thermos with a gloved hand. He'll extend the semen so that one ejaculation can make about twenty doses of semen. Once again, it's an economic thing. It gives us more use of that one boar, instead of having to feed so many boars. But it goes beyond that. We can change our genetics faster than we could if we had a stable of twenty or twenty-five boars. It's better to have one really good boar and use his semen."

Next stop is the nursery—that's the industry term for a place where early-weaned pigs are given special feed to enable them to survive the stress of separation and weaning. Each of the pens contains a few dozen small pigs. We ask him how the pigs handle the stress of weaning. "Oh, there really isn't much that happens. The first day they just kind of lay around. Whenever you come in the building, they grunt and make a lot of noise because they are used to having mama around."

We get back in the truck and go to another farm where Wayne has a finishing building, where the pigs are grown to market weight. Along the way, we talk about the changes in farming we've both seen over the years. He mentions the loss of middle-income people in the rural areas around him. Now, he says, "We've got a bunch of people that are looking for $150-a-month houses to move into. They're making meth and they're making trouble. The rural countryside has changed dramatically." (Making methamphetamine, in Iowa? At the time, we thought Wayne must have been exaggerating, but when we checked it out, we found that Iowa has the second highest number of meth labs and the fourth highest level of meth use in the nation.[19] Is that a consequence of the loss of family farmers too, we wondered?)

The conversation drifts to the price of corn and subsidies. Wayne thinks that years of government subsidies have kept corn artificially cheap for livestock producers. "We've been producing grain below the cost of production for so many years that it's just a given. It's a guarantee. If we'd gotten corn prices up to where they ought to be, a lot of this livestock thing never would have happened. It's been on the back of cheap grain. I don't know how you change that."

We have reached the finishing building. It's open on both sides. Running the length of the building on each side is a plastic curtain he can roll up and down to adjust temperature and ventilation. "Let's open the door just to give you a whiff of the air quality." We step through to the pens. On this mild spring day, a breeze is blowing over the pens full of pigs. It is total confinement, but with a breath of semi-fresh air. "If it's thirty below zero outside, these curtains will be closed and the furnace will be running a bit. These old pigs here'll be all stretched out and as comfortable as if they were in the Bahamas."

We ask him about drugs and medications administered to these pigs. He says that this is "one of those deals that gets misrepresented. People think we're feeding a lot of antibiotics out here. Our whole goal is not to feed a lot of them because they cost us money." He explains that when he first brings pigs into a finishing building he gives them a dose of the antibiotic tetracycline in their feed "just to give them something for the stress in moving them." Then he puts them on "a growth promotant called BMD, bacitracin something something. I can't tell you what all is in it. It helps them grow faster and that's the name of the game." (He's referring to Bacitracin methylene disalicylate, another antibiotic.) If the pigs develop diarrhea or "a cough or a problem," they give them an antibiotic or other medication, usually in the water. "A pig will drink when they won't eat," he explains.

Our tour is over and it's time to leave. We're sitting in Wayne's pickup truck in the driveway back at his home. He emphasizes again how he wants to get the right story out there. "What really concerns us in animal agriculture is that we've been made out to be the bad guys. We're working hard to produce a quality product and we're treated like we're just... well, terrible people. It doesn't go down well in the ag community."

We don't think he's a bad guy. He worked hard to buy the farm from his parents and brothers—a farm that had been in the family for a century—and he found a way to keep it going when most family pig farms were going out of business. We like the fact that he doesn't keep his pregnant sows in crates—probably the least defensible aspect of standard pig confinement practices in America. We particularly admire his openness about what he is doing, a refreshing contrast to all the other intensive pig producers we contacted. His method of disposing of his manure seems more responsible than that of many pig producers. We appreciate that his buildings keep his pigs warm in the cold Iowa winters. But we wondered if there couldn't be a way of keeping them warm and giving them a better life than they can have living in an environment as barren and restrictive as his total confinement buildings.

When we sent Wayne what we had written about our visit, in addition to asking not to be identified and making a few other minor suggestions, he and Mrs. Bradley wrote that they thought our final sentence—the one you have just read—made "no sense." Instead they suggested a different way of ending our account of our visit to their farm. Here it is, with their original underlining. You be the judge.

> Raising pigs today is so much improved over methods used by our great-grandparents, <u>and the meat that we consume is so much leaner and healthier for us to eat</u>! The highest-quality standard of 'the other white meat' is the goal of USA pork production in the 21st century. Let's thank the American farmer for a solid science-based industry that includes good animal care while being good stewards of the environment. Let's enjoy that pork chop hot off the grill, or that pork roast with potatoes and carrots, <u>because there's no safer food source than USA-raised pigs</u> for the pork consumers in the USA and other countries which import our pork!

## PROFITABILITY AND ANIMAL WELFARE

The real ethical issue about factory farming's treatment of animals isn't whether the producers are good or bad guys, but that the system seems to recognize animal suffering only when it interferes with profitability. The animal industry always says that producers take care of their animals because what is good for the animals is good for the producer. Professor Bernard Rollin, who has taught veterinary ethics at Colorado State University for almost thirty years, has given a graphic example of how profitability and animal welfare can pull in opposite directions. A veterinarian was visiting a 500-sow, "farrow to finish" swine operation with three full-time employees and a manager. He noticed that one of the sows in the gestation crates had a hind leg sticking out at an odd angle. When he inquired, he was told "She broke her leg yesterday, and she's due to farrow next week. We'll let her farrow in here, and then we'll shoot her and foster off her pigs." The vet was troubled by the idea of leaving the sow for a week with a broken leg and offered to put the leg in a splint, charging only the cost of his materials. He was told that the operation could not afford the manpower involved in separating and caring for the sow. At this point, the vet, who had been brought up on a family pig farm where the animals had names and were treated as individuals, realized that "confinement agriculture had gone too far."[20]

Is this an extreme case, or common practice? The cost calculations that Wayne made when discussing the possible use of a local anaesthetic to reduce the pain of operations like castration—"there's not a dollar a pig to throw away"—show that this kind of thinking is built into intensive animal raising. As long as the market provides no incentive for reducing the pigs' pain, the pig producer cannot afford to spend more than a penny, or perhaps a nickel, for that purpose. If he does, someone else who won't spend anything to reduce pain will produce cheaper pigs and put him out of business. That is why the way that factory farming treats animals is not so much a problem of gratuitous cruelty or sadism, and the main problem is not a matter of preventing isolated incidents of animal abuse. The core issue is the commercial pressures that exist in a competitive market system in which animals are items of property, and the conditions in which they are kept are not regulated by federal or state animal welfare law.

## TRACKING DOWN JAKE'S MILK

Jake thought that she was buying milk from local farms because Coleman Dairy, the brand she bought at Wal-Mart, is an Arkansas-based corporation. When we called and asked if we could see their cows, however, Walt Coleman told us that they hadn't had any cows since 1935. They buy their milk from Dairy Farmers of America, a big dairy cooperative, and although some of their milk comes from Arkansas, it can also come from Texas or New Mexico. Coleman wasn't willing to help us any further in our quest to see the source of Jake's milk.

Milk and cheese production enjoy a better reputation than other forms of intensive farming, and the dairy industry is keen to keep it that way. In advertisements for dairy products, it's common to see cows enjoying acres of rolling green pasture, often with their calves nearby. The impression many consumers get is that dairy cows lead natural lives, and we humans merely take the surplus milk that the calf does not require. People also think that cows are placid animals without much of an emotional life. Both are misconceptions. Cows have strong emotional lives. They form friendships with two, three, or four other cows, and, if permitted, will spend most of their time together, often licking and grooming each other. On the other hand, they can form dislikes to other cows and bear grudges for months or even years.

More remarkably still, cows can get excited when they solve intellectual challenges. Donald Broom, professor of animal welfare at Cambridge University, set cows a problem—to work out how to open a door to get some food—while measuring their brainwave patterns. When the cows solved the problem, Broom reported, "Their brainwaves showed their excitement; their heartbeat went up and some even jumped into the air. We called it their Eureka moment."[21]

Peter Lovenheim is a writer who lives in Rochester, New York. He was standing in line at McDonald's one day when he decided that he'd like to know more about how a hamburger is produced. He bought three newborn calves and had them raised in the usual way until it was time to slaughter them. Because Rochester is close to many of New York State's dairy farms—and New York is the third largest dairying state in the U.S., after Wisconsin and California—Lovenheim bought male calves from a nearby dairy farm. Most males born to dairy calves are raised for veal, or slaughtered immediately for pet food, but a few of the stronger ones are raised for beef. Thanks to Andrew and Sue Smith, who were remarkably open about what they do, Lovenheim was able to spend a lot of time at Lawnel Farm, and the following account draws on his description of Lawnel when he was there in 2000.[22]

With about 900 cows being milked—that doesn't include young cows who were not yet giving milk, nor cows who were temporarily not lactating—Lawnel was a medium-sized dairy operation, larger than some of the organic farms we will describe in Part II but small compared to, say, Bill Braum's dairy near Tuttle, Oklahoma, which milks over 10,000 cows, or Threemile Canyon in Oregon, which milks 18,000 cows.[23] A Cornell University study expects the number of dairy farms in the United States to decline from 105,000 in 2000 to 16,000 in 2020, while the number of cows per farm and the total milk production both increase.[24]

At Lawnel, the cows were kept indoors, in barns. Unlike many dairy farms, they were free to walk around inside the barn—they were not in "tie-stalls" that confine cows, for most of the year, to a single stall where they are fed and milked. In the western United States, dairy cows are more likely to be kept outside, but even then they are just in dirt lots. Very few dairy cows in the U.S. get to graze in the grassy meadows typical of dairy industry advertising—the exceptions are mostly cows producing milk certified "organic," but, as we shall see, even some of them are not on pasture.

The modern dairy cow has been bred to produce as much milk as possible and now produces more than three times as much milk as a typical dairy cow did fifty years ago.[25] The result is considerable stress on the cow's body. To increase milk production still further, the Smiths gave their cows injections, every other week, of BST, or bovine somatotrophin, a genetically engineered growth hormone. BST is banned in Canada and in the European Union because of concerns for the health and welfare of dairy cows, but it is widely used in the United States. It increases milk production by about 10 percent, but the site of the injection may become swollen and tender. BST can also increase problems with mastitis, a painful udder infection that afflicts about one in six U.S. dairy cows.[26] Sue Smith said she didn't like giving the injections, but "If we're making more milk and it's profitable, it's something we should be doing."[27]

Like human females, dairy cows do not give milk until they have given birth, and their milk production will begin to decline some six months after the birth. So after they reach maturity they are made pregnant by artificial insemination roughly every year. Normally a calf would suckle from its mother for six months, and the bond between mother and child would remain strong during that period, but dairy farms are in business to sell milk, not give it to calves. At Lawnel Farms, Lovenheim watched a cow give birth and begin to lick her calf, but forty minutes later a farmhand came and took the calf away. The cow sniffed the straw where the calf had been, bellowed, and began to pace around. Hours later she was sticking her nose under the gate to the barn in which she was confined, bellowing continuously. Meanwhile her calf was in another part of the farm, lying shivering on a concrete floor. Within a few days he was dead, and his body was lying on the farm's compost pile.[28]

Oliver Sacks, who writes about people with unusual neurological conditions, spent some time with Temple Grandin, the livestock consultant McDonald's has employed to advise them on animal welfare issues. Sacks was more interested in Grandin's autism than in her work with animals, but he accompanied her on a visit to a dairy farm. As Sacks describes it: "We saw one cow outside the stockade, roaming, looking for her calf, and bellowing. 'That's not a happy cow,' Temple said. 'That's one sad, unhappy, upset cow. She wants her baby. Bellowing for it, hunting for it. She'll forget for a while, then start again. It's like grieving, mourning—not much written about it. People don't like to allow them thoughts or feelings.'"[29] John Avizienius, the senior scientific officer in the Farm Animals Department of the RSPCA in Britain, says that he "remembers one particular cow who appeared to be deeply affected by the separation from her calf for a period of at least six weeks. When the calf was first removed, she was in acute grief; she stood outside the pen where she had last seen her calf and bellowed for her offspring for hours. She would only move when forced to do so. Even after six weeks, the mother would gaze at the pen where she last saw her calf and sometimes wait momentarily outside of the pen. It was almost as if her spirit had been broken and all she could do was to make token gestures to see if her calf would still be there."[30]

Female dairy calves may be reared as replacements for the "culled" cows who get sent to slaughter. Although the natural lifespan of a cow is around 20 years, dairy cows are usually killed at between five and seven years of age, because they cannot sustain the unnaturally high rate of milk production. Male calves who survive are sent to auction at an age when they can barely walk. Temple Grandin has strong views about that, too: "Worst thing you can do is put a bawling baby on a trailer. It's just an awful thing to do."[31]

The usual options for these male dairy calves are, as already mentioned, to be slaughtered immediately or to be raised for "milk-fed" veal. From the calf's point of view, immediate slaughter is the better fate, for it spares him 16 weeks of confinement in semi-darkness, in a bare wooden crate too narrow to turn around. He will be tied at the neck, further restricting his movements. Already stressed by separation from his mother and unable to mingle with others of his kind, he will be fed only "milk replacer," a liquid mixture of dried milk products, starch, fats, sugar, antibiotics, and other additives. This diet is deliberately so low in iron that he will develop subclinical anemia. That's what the veal producer wants, because it means that the calf's flesh, instead of becoming the normal healthy red color of a 16-week-old calf on pasture, will retain the pale pink color and soft texture of "prime veal." Bought mostly by expensive restaurants catering to gourmet tastes, that kind of veal fetches the highest price. For the same reason, the calf will be denied hay or straw for bedding—if he had it, his desire for roughage and something to chew on would cause him to eat it, and since it contains iron, that too would change the color of his flesh. The wooden stalls and neck tether are part of the same plan. If the stall had iron fittings, he would lick them, and if he were able to turn around, he would lick his own urine—again, in order to satisfy his craving for iron.

Apart from the separation of cows from their calves and the way the newborn male calves are treated, the most disturbing passages in Lovenheim's description of Lawnel Farm portray the treatment of "downers"—cows who, through illness or accident, are no longer able to stand. On one occasion Lovenheim saw Sue Smith trying to raise a downed cow, No. 4482. She started by coaxing her with gentle words, but when that didn't work she twisted the stump of the cow's tail, then jabbed her knee into the cow's side and screamed into her ear. When that met with no success she twisted the cow's ear and jabbed her several times in the ribs with an electric prod. That didn't work either. If a downed cow can't be raised, she is dragged out. So they called Bill, a renderer, to take the cow away. Lovenheim describes what happened next: "Andrew gets on a small tractor and backs it through the barn door while Bill ties a sling around 4482's front right hoof. When the sling is attached to the tractor, Andrew reverses direction, dragging the downed cow thirty or forty feet across the barn door, her useless back feet spread wide, her left front hoof kind of paddling along to keep up." (While this was going on, Andrew and Bill discussed what crops the farm had planted that season.) Once out of the barn, Bill winched the cow up a steeply inclined ramp into the back of a truck that took her to the slaughterhouse. After watching this, Lovenheim asked Sue if she had ever considered euthanizing downed animals on the farm. She told him that they'd done that once, but the procedure was expensive.[32]

Manure from dairies, like that from chicken and pig factory farms, pollutes rivers, kills fish, and ruins the homes of nearby residents.

Another pollution problem, more specific to cows, is often treated as a joke. When cows ruminate, or "chew the cud," they produce gases called "volatile organic compounds." (For those who like anatomical details, most is generated in burps rather than farts.) When there are a lot of cows, that makes for a lot of gas and can cease to be a joke. The San Joaquin Valley, part of California's Central Valley and one of the world's richest agricultural regions, ranks alongside Houston and Los Angeles as having the worst air pollution in the United States. Over the last six years, the valley has violated the federal limit on ozone smog over an eight-hour period more often than any other region in the country. Officials from the San Joaquin Valley Air Pollution Control District believe that the valley's 2.5 million dairy cows are the biggest single source of a major smog-causing pollutant and are trying to force the dairy industry to do something about it. Other gases are emitted by cow manure and the lagoons in which it is stored. The dairy industry is resisting proposals for change. Tom Frantz, who says he has developed asthma as a result of dairy farms moving near to him, heads a group called the Association of Irritated Residents that is calling for stricter regulation. Frantz says: "Ag hasn't been regulated in the past, but times are changing. Our lungs will not become an agricultural subsidy."[33]

The problem isn't only one for local residents, either. The gases contain methane, which contributes significantly to global warming. In that respect we are all subsidizing agriculture.

## THE BEEF INDUSTRY

By a curious coincidence, about the same time that Peter Lovenheim in New York was buying calves in order to follow the process of turning a calf into a hamburger, Michael Pollan, another writer, was doing much the same thing in the Midwest. Lovenheim's calves were byproducts of the dairy industry and were raised by a dairy farmworker and his wife who kept about a dozen cattle on the side. Pollan bought a young steer—a castrated male—from a ranch in South Dakota and had him fattened alongside 37,000 other cattle on a feedlot in Kansas. The dairy industry that Lovenheim observed is the source of about half of the hamburger meat served at the fast-food restaurants Jake likes to frequent. The beef industry that Pollan portrays is where most of America's 36 million beef cattle are produced every year and is the likely source of the porterhouse steak she buys at Wal-Mart.

Pollan's calf, known as 534, was born in March. The calf remained with his mother for more than six months, part of a herd that had many acres of prairie pasture on which to graze. He wasn't even weaned until October. But it was all downhill from there. The young steer was loaded into a truck and driven 500 miles to Pokey Feeders where, in Pollan's words, "Cattle pens stretch to the horizon, each one home to 150 animals standing dully or lying around in a grayish mud that, it eventually dawns on you, isn't mud at all." When Pollan visited, he could smell a "bus-station-men's-room" odor more than a mile before he got there. Here 534 lived another eight months, until slaughter.[34]

On arrival at the feedlot, 534 was given an implant of a synthetic hormone in the back of his ear—something similar to the muscle-building testosterone surrogates that athletes use. Giving them to cattle is banned in Europe because of concerns about the potential health risk of drug residues, and of course U.S. law prohibits people from self-medicating with steroids. In the U.S., however, giving them to cattle is standard practice. It makes them put on more muscle, which means more money for the growers. When Pollan asked Rich Blair, the rancher from whom he bought 534, what he thought about the hormone implants, Blair said: "I'd love to give up hormones. The cattle could get along better without them. But the market signal's not there, and as long as my competitor's doing it, I've got to do it, too."

Instead of grass, 534 now ate corn kernels, together with a daily dose of antibiotics to enable him to survive on this diet. Dr Mel Metzen, the staff veterinarian at Pokey Feeders, told Pollan that a great many of the health problems that he and his eight assistants have to deal with stem from the diet. "They're made to eat forage," Metzen says, "and we're making them eat grain." Ruminant animals have a digestive system that has evolved to break down grass. If they don't get enough roughage, they develop lactic acid in their rumens, which creates gas and causes "feedlot bloat," a condition so severe that cattle can suffocate from it. Liver abscesses are also frequent. Putting cattle on a corn-based diet is like putting humans on a diet of candy bars—you can live on it for a while, but eventually you are going to get sick. For the beef producer that doesn't matter, as long as the animal doesn't drop dead before being slaughtered. By feeding antibiotics on a daily basis, the risk of that happening is reduced to manageable proportions—and it is a risk worth taking, because the cattle reach market weight in fourteen months, rather than the eighteen months to two years they would otherwise take. Without antibiotics, Metzen admitted, it wouldn't be possible to fatten cattle on corn. "Hell, if you gave them lots of grass and space," he joked, "I wouldn't have a job."

Corn isn't the only strange food that cattle are fed. When mad cow disease became a major issue in Europe, the public was surprised to learn that it was caused by cattle eating the remains of sheep who had been infected with a related disease. Since when, people asked, do cattle eat meat? In fact, slaughterhouse leftovers have been going into cattle feed for about forty years, because they are cheap and add protein to the diet. In the wake of the mad cow disaster, most countries placed restrictions on feeding meat remnants to cattle, but in the U.S. it is still, at the time of writing, permitted for cattle feed to contain beef blood and fat, as well as gelatin, "plate waste"—restaurant leftovers—chicken and pig meat, and chicken litter—which includes fecal matter, dead birds, chicken feathers, and spilled feed. The spilled feed can include the same beef and bone meal that is not allowed to be fed to cattle directly, but can be fed to chickens.

In January 2004 the Food and Drug Administration announced plans to ban blood, plate waste, and chicken litter, and an international review panel convened by the Secretary of Agriculture recommended banning all slaughterhouse remnants, two years later, none of these proposed bans had come into effect. Frustrated at the delay, scientists and McDonald's Corporation told the FDA that stronger steps were needed to stop mad cow disease, which the researchers called "an insidious threat." McDonald's Vice-President Dick Crawford called on the government "to take further action to reduce this risk."

One of the reasons for the delay, according to Stephen Sundlof, Director of the FDA's Center for Veterinary Medicine, was that the proposed ban on the use of chicken litter generated "huge concern" from chicken producers. No wonder—about a million tons of chicken litter are disposed of by being fed to cattle each year. That means that, on average, each of the 36 million cattle produced in the U.S. has eaten 66 pounds of it. In other words, environmental problems created by the chicken industry are preventing the FDA from taking steps recommended by public health experts to ensure the safety of U.S. beef.

The feedlot system is also an ecological disaster. When we eat ruminants who have been grazing on pasture, we are, in effect, harvesting the free energy of the sun. But feedlots thrive because in the U.S. bulk corn sells for about 4 cents a pound—less than the cost of production, thanks to the billions of taxpayers' dollars the government gives in subsidies to the growers. (Most of the cash goes to people who are already very wealthy.) The corn in turn requires chemical fertilizers, which are made from oil. So a corn-fattened feedlot steer is, as Pollan says, "the very last thing we need: a fossil-fuel machine." Pollan asked David Pimentel, a Cornell ecologist, to calculate how much oil went into fattening 534 to his slaughter weight of 1,250 pounds. Pimentel's answer: 284 gallons.

Then there is the issue of what happens to the run-off from the feedlots. Nebraska is *the* state for big feedlots, with 760 of them authorized to have more than 1000 head of cattle. The largest, near Broken Bow, is licensed for 85,000 cattle. Alan Kolok, a professor of biology at the University of Nebraska, is studying the impact of feedlots on streams that flow into the Elkhorn River. We met him in Omaha, and he drove us west into Cuming County, one of the nation's top beef-producing counties. Near West Point, we came to a feedlot for about 5,000 cattle—the usual fenced, bare, dirt-and-manure yards with bored-looking cattle standing in the sun. It was June, and although it wasn't especially hot yet, we remarked on the lack of shade, saying to Alan that if any had been provided, most of these cattle would have been standing in it. He said that the weather was going to get hotter. Indeed, by the end of July, much of Nebraska had had 30 days with temperatures above 90 degrees, and several above 100. Roxanne Bergman, who runs a "dead-stock removal" company in Clearwater, said her company alone had hauled out 1,250 dead cattle during a few days of hot weather and could not handle all the calls it was receiving.[36]

Researchers from the Department of Animal Science and Food Technology at Texas Tech University studied the use of shade in feedlots. The study divided cattle into a group that had shade available and one that did not have shade available. The cattle with shade available "used the shade extensively" from 9.00 A.M. to 5.30 P.M., following the shade as the sun moved. Cattle without shade were four times as aggressive to other cattle than those with shade. But the researchers also noted that "In west Texas, shade is generally not used in commercial feedlots because it is not thought to be cost effective."[37] Once again—and not only in west Texas—when better animal welfare costs money, animal welfare loses.

Alan showed us how the feedlot we were looking at had been built right down to the edge of the north fork of Fisher Creek. A holding lagoon built to catch the feedlot run-off, filled with unpleasant-looking brown water, was separated from the creek by an earth embankment. Alan explained that in heavy rain, it was likely that polluted water would run off from the feedlot into the creek, or could seep through the embankment into the creek. We drove on and came to another feedlot on sloping land not far from the Elkhorn River. Here Alan has found local fish, fathead minnows, showing signs of altered sexual features. As compared with fish captured near a wildlife refuge where there are no feedlots, the male minnows had less pronounced masculine features and females had less pronounced feminine features. This phenomenon is known as "endocrine disruption." If fathead minnows are altered, the same could happen to fish used for recreational fishing, like bass and catfish, and the Nebraska Department of Game and Parks is concerned about the problem. Alan and his colleagues have published studies hypothesizing that the most likely explanation is the steroids implanted in the feedlot cattle. The cattle excrete them, and when it rains they wash off into the rivers, where they have a half-life of 6 to 12 months.[38]

Although Nebraska livestock producers say that their state has some of the strictest regulations in the nation, there is very little enforcement of regulations regarding feedlots. In addition to its 4,560 cattle feedlots, Nebraska has thousands of confined pig units, and of course egg and chicken producers as well. In 1999, the Nebraska Department of Environmental Quality stated that there were 25,000 to 30,000 hog and cattle feeding operations in the state, most of which had never applied for permits from the agency, although state law had required permits since 1972. Even if these pig farms and cattle feedlots had applied for permits, the Department simply would not have had the staff to inspect more than a small fraction of them. In 1997, the Department's director testified that he had a staff of five for issuing permits and inspecting livestock-feeding operations and that they "tried" to inspect 225 of the larger operations.[39]

It's not unusual in the U.S. for state departments to lack the resources to monitor water pollution. Idaho, it seems, is in a similar position to Nebraska. Mike Bussell, director of the Environmental Protection Agency's regional office of compliance and enforcement, said that his office was going to have to start inspecting feedlots in Idaho because the Idaho State Department of Agriculture was "never able to accomplish" the basic task of producing an "overall inventory of the regulated community, so we'd know how many operations we were dealing with, and who needs to comply."[40] In Michigan, according to a regional Environmental Protection Agency report, the Department of Environmental Quality "does not conduct inspections to determine compliance by CAFOs (Concentrated Animal Feeding Operations) with permit application and other program requirements."[41]

If the untreated waste from feedlots doesn't flow directly into the streams and rivers, it will be sprayed onto fields through a center-pivot irrigation system. Manure is wet and costly to transport, so it is spread on fields close to the animal feeding operations, often in quantities too great for the soil to absorb, and in heavy rain, it runs off into the creeks. (The method of working it into the soil used by Wayne Bradley in Iowa isn't widespread because it takes more labor.) In 2002, the Nebraska Department of Environmental Quality sampled about 5,000 of the state's more than 16,000 miles of rivers and streams and found that pollution exceeded the standard for uses like recreation, aquatic life, agriculture, and drinking supply in 71 percent—a significant jump on the already alarming 58 percent found to be polluted in 2000.[42] Dennis Schueth, who manages the Upper Elkhorn Natural Resources District, told Nebraska Public Radio Network reporter and author Carolyn Johnsen: "We can be more environmentally sound if we want to pay more for our food."[43] Right. But what mechanism is supposed to bring about that outcome? Even if Jake and Lee were willing to pay more for their meat in order to protect the environment in Nebraska, how could they be sure—or even reasonably hopeful—that the extra dollars they were spending were having this effect? In later chapters we'll consider some possibilities.

As we drove back along Route 275 near West Point, Alan pointed out dozens of big containers of anhydrous ammonia—a synthetic nitrogen fertilizer. "Isn't it odd," he asked, "that all this synthetic fertilizer is being used in the midst of a feedlot region, where there is all this much better natural fertilizer available?"

## AUSTRALIAN BEEF

Raising beef doesn't have to be like this. On a visit to Australia, we met Patrick Francis, the editor of *Australian Farm Journal,* a popular farming magazine. Patrick had heard of our interest in ethical farming and invited us to look at the small beef property—in America it would be called a ranch-that he ran with his wife Anne near Romsey, in Victoria. The property was a delight to stroll around, in part because Patrick and Anne have set aside 20 percent of the land for revegetation, mostly with native eucalypts. The straightest trees will, in time, be sold for timber, but meanwhile, by storing carbon, they are making a small contribution to mitigating global warming. A recent carbon balance calculation showed that each year the farm was absorbing from the atmosphere 220 tons more carbon dioxide than it was emitting. The plantations also provide habitat for native animals, including a mob of gray kangaroos who hopped away as we strolled by. Meanwhile, in the open fields, the cattle made a remarkable contrast to the dusty, manure-caked animals we saw in the bare Nebraska feedlots.

It was mid-April, the southern hemisphere autumn, and there had been little rain for months, but Patrick rotates his cattle around different fields every week or two, a technique that gives the grass time to recover from grazing and ensures healthy soils and well-grassed pastures. This method eliminates the need to conserve fodder—Australian winters are mild and free of snow, and there is enough grass in the fields for the cattle to eat all year round. The rotation makes for thick pastures, which eliminates the need to use pesticides for weed control.

The day we were there coincided with one of those rotations, and we watched as Patrick moved the cattle on to the next field. He has a way of calling his cattle to him, and they follow where he leads. First among them is a particularly affectionate seven-year-old bullock—a term used for an older steer—who Patrick has kept on the farm for his leadership role in showing the newer cattle on the property what to do. (His flesh, by now, would be too tough for anything but hamburger.) The day was pleasant, and the sun had lost the sting it has at the height of summer, but once the cattle had moved into the new pasture, they soon found the shade cast by a row of cypresses, and most of them stood under the trees. Though the youngest calves were six months old, they were still keeping company with their mothers. The lives of these cattle were, it seemed, entirely comfortable. They had what cattle need: plenty of grass, clean water, shade, and their own social group.

Patrick told us that he prefers to sell his cattle direct from the farm to the slaughterhouse, but there are times of the year when he doesn't have enough grass on his pasture to get them ready to market. Then he sells them to a feedlot for short-term fattening. For the Australian domestic market, only about 25 percent of cattle are fattened in feedlots, although that percentage is growing because supermarkets prefer the greater reliability of the quality of the meat. Nevertheless, most Australian cattle are fed for only 70 days, less than half the normal period in feedlots in the U.S. For export markets—predominantly Japan and Korea, with a small amount going to the United States—cattle are generally fed for about 150 days, because consumers there have developed a taste for the marbled, fattier meat that results from fattening cattle largely on grain for a longer period of time.

## SLAUGHTER

Mammals killed for food in the U.S.—unlike chickens, ducks, and turkeys—are required by law to be stunned before being killed. No, that's not quite right: the U.S. Department of Agriculture ludicrously classifies rabbits as poultry, although they are mammals, thus allowing producers to avoid the legal requirement to stun them before slaughter. Temple Grandin surveyed American slaughterhouses to find out what percentages of animals are rendered insensible by the first application of the stun-gun. In her first survey, in 1996, only 36 percent of slaughterhouses were able to effectively stun at least 95 percent of animals on the first attempt. Six years later, 94 percent were able to do so. That is a dramatic improvement, and, as we will see in the next chapter, there is a reason for it.

Nevertheless, as a General Accounting Office report to Congress on the enforcement of the Humane Methods of Slaughter Act acknowledges, despite the improvement, setting a standard of only 95 percent of animals being stunned on the first attempt "still indicates that hundreds of thousands of animals were not stunned on the first try . . . Thus, there may be undetected instances of inhumane treatment." The report notes that there were "approximately six observations for HMSA compliance per month, or less than two observations per week, for each of the 918 plants that are covered by the act." In other words, with hundreds of animals being killed every hour, the inspectors are rarely present. When they are there, the plant operator knows it, and so what the inspectors observe may not be representative of what happens when they are absent. Even when inspectors are present and do find violations, the report found that enforcement polices were inconsistent and "inspectors often do not take enforcement action when they should."[44]

A video taken by an undercover investigator at AgriProcessors, Inc, in Postville, Iowa, during the summer of 2004 shows what can happen when inspectors are not present. AgriProcessors, Inc. is a kosher slaughterhouse, which means that it kills animals in accordance with orthodox Jewish dietary law, which forbids stunning before slaughter. In theory, in kosher slaughter animals should be killed quickly and cleanly by having their throats cut with a single slash of a sharp knife. Unconsciousness from loss of blood to the brain should follow within a few seconds. In the video, however, cattle who have had their throats cut and their tracheas removed still thrash around for a long time before they die. Some struggle to get to their feet—and even succeed in standing up. While this happens, a worker waits for the animal to collapse so that he can tie a chain around its rear leg and hoist it off the ground. One animal goes so far as to stagger through an opening into a different area of the slaughterhouse before collapsing. Two more cattle come down the killing line and have their throats cut before this one is finally hoisted off its feet and dragged away.[45]

We are not suggesting that these scenes are typical of kosher slaughter, or of American slaughter in general. But it is worth noting that AgriProcessors is the world's largest kosher slaughterhouse, and its owner has stated that "What you see on the video is not out of the ordinary." Similarly, the Orthodox Union, the world's largest kosher certifier, has defended the plant consistently and has said that the plant meets "the highest standards of Jewish law and tradition" and that its kosher status has never been in jeopardy.[46]

Since inspectors are not assigned to the point of kill in any U.S. slaughterhouses, it is probable that anyone who eats meat will, unknowingly, from time to time be eating meat that comes from an animal who died an agonizing death.

# 5

# CAN BIGGER GET BETTER?

It's not surprising that Jake likes McDonald's—so many people do that it has become the world's largest restaurant chain, with 31,000 restaurants in 119 countries. But it is also not surprising that when French farmer José Bové wanted to protest against globalization and the Americanization of French culture, he chose a McDonald's restaurant to drive his tractor through. The Golden Arches have become a symbol of America—and to many, a symbol of everything that is wrong with America. Food, architecture, cultural or economic imperialism—you name it, if it is American and you don't like it, you can protest about it at your local McDonald's.

Reality, as always, is more complicated. Consider McDonald's record on animal welfare. Ray Kroc, McDonald's founder, had a reputation for implacably tough management. "What do you do when your competitor is drowning?" he is said to have asked, then answering himself: "Get a live hose and stick it in his mouth."[1] So when, in the early 1990s, a New York animal activist named Henry Spira decided to try to pressure McDonald's into developing less inhumane ways of raising the animals, he knew it wasn't going to be easy. Spira was essentially a one-man-band, running an organization called Animal Rights International that had no paid staff and no office except the modest, rent-controlled New York apartment in which Spira lived. But Spira knew what it was like to take on the big boys. He had been involved in the civil rights movements in the South in the 1950s and 1960s. As a sailor in the merchant marines, he had been part of a reform group battling a corrupt union boss who was not averse to hiring thugs to beat up his opponents. He took up the cause of animals in his late forties and applied the lessons he had learned in his earlier struggles.

After several startling successes in the field of research on animals, Spira turned his attention to farm animals. He bought shares in McDonald's and took steps to move a resolution at their 1994 shareholders' meeting on the treatment of the animals they served. McDonald's persuaded him to withdraw the resolution, pledging to issue a statement requiring their suppliers to treat animals humanely. Within the animal movement, many thought the statement was meaningless and that Spira had been bought off too easily. Spira's view was that while the statement itself might not mean a lot, it was a first step in setting an industry standard. "If McDonald's moves a millimeter," he said, "everyone else moves with them." Privately, though, Spira was far from satisfied. He wanted verifiable standards for humane handling and slaughter, and he wanted McDonald's to pledge to cut off suppliers who did not meet those standards. On that, McDonald's stalled. It was only after the public relations disaster of the McLibel trial in 1997 that McDonald's attitude changed.

We have already mentioned some of the findings of the McLibel trial. Three months before the McLibel verdict, McDonald's Vice-President Shelby Yastrow had bluntly told Spira that "farm animal well-being is not high on McDonald's priority list." When Spira called immediately after the verdict, Yastrow's interest in farm animal well-being had jumped enough for him to fly to New York to talk about it. At that meeting, Spira pressed Yastrow to meet livestock consultant Temple Grandin to discuss some practical and feasible improvements to the handling of animals slaughtered for McDonald's hamburgers. Yastrow, who was now also coming under increasing pressure from People for the Ethical Treatment of Animals, agreed to do so. Soon afterwards, McDonald's commissioned Grandin to develop an animal welfare auditing system for the slaughterhouses that supply it with meat. By 1999, Grandin's audit system was up and running, and McDonald's told suppliers that failed the audit to improve their performance within 30 days or lose McDonald's as a customer. Three years later, Grandin said that she had seen more change in slaughterhouses since 1999 than in all her previous 30 years in the field.[2] McDonald's is, of course, a global enterprise. Although it does not set the same standards for every market in which it operates, Grandin has worked with McDonald's suppliers in Australia, Brazil, Britain, China, New Zealand, Norway, and Thailand. She has visited Australia three times, auditing the facilities of McDonald's suppliers of beef, pork, chicken and eggs. She also holds training days for the staff of the corporations that supply McDonald's, and has set up an independent third-party auditing system to check that the suppliers are operating satisfactorily. Simone Hoyle, who is Director of Supply Chain Management for McDonald's Australia, claims that these visits have led to "significant improvements" since Grandin's first visit in 1999.[3]

The changes didn't stop at the slaughterhouses. In August 2000, McDonald's shocked the egg industry by announcing that it would require its U.S. egg suppliers to increase the amount of space they gave to hens by almost 50 percent, up from an industry average of 50 square inches per bird to a minimum of 72 square inches. Furthermore, McDonald's would no longer accept eggs from birds who had been "force molted"—the practice described earlier of starving hens to jumpstart and synchronize the period for which they will lay eggs. McDonald's also said it would seek to phase out the painful practice of beak searing. At that time neither United Egg Producers nor any other U.S. body had any animal welfare standards for how hens should be kept. The fast food chain was, according to Bob Langert, McDonald's director of environmental affairs, planning to be a leader on these issues.[4]

And indeed, given the U.S. egg industry's zero animal welfare baseline, McDonald's was leading the way. PETA now switched its attention to the second largest fast-food chain, Burger King, and they soon announced their own similar set of standards. Next it was Wendy's turn. At this stage, rather than have its members picked off one by one in order of sales, the National Council of Chain Restaurants got together with the Food Marketing Institute to set standards for all their members. Together, the membership of these organizations operates three-quarters of all food retail stores in the U.S. and owns or franchises 120,000 restaurants.[5] Though he had not lived to see it, Spira was proved correct—McDonald's had moved a millimeter—and then kept moving—and the rest of the industry did follow.

## A LESS CRUEL BIG MAC?

The changes McDonald's made need to be put in perspective. On the handling and stunning of cattle at the slaughterhouse, Grandin has said: "There's a pre-McDonald's era and a post-McDonald's era. It's as different as night and day."[6] Not everyone is so sure the audits can be trusted. If they are scheduled in advance, it is possible for managers to run the killing line more slowly than usual during the audit in order to miss stunning fewer animals and so meet the required standards. Even with unannounced audits, there are stories told of warning systems set up to give immediate notice when the auditors arrive in the parking lot, so that the line can be slowed before they reach the killing room. Nevertheless, we are prepared to accept Grandin's judgment that, as a result of McDonald's actions, there has been a dramatic improvement in handling cattle and pigs at slaughter. Unfortunately, McDonald's has done nothing to improve the conditions of slaughter for the chickens they serve.

Stopping the prolonged cruelty of starving hens to make them molt is also important. The value of the additional space McDonald's requires for hens is more debatable. It merely brings these birds up to a minimum that the European Union is phasing out, after receiving scientific veterinary advice that it is inadequate. Although McDonald's appointed Professor Joy Mench to their Animal Welfare Council, they remain unmoved by her judgment that even 80 inches would not be enough for a hen to be able to perform the normal behavior that enables her to feel comfortable.[7] Diane Halverson, another member of McDonald's Animal Welfare Council and a farm animal adviser for the Animal Welfare Institute, is troubled by the idea that the public might think of these standards as really ensuring the welfare of hens. "These are not animal-welfare standards," she told a reporter for the *San Francisco Chronicle,* "they are standards of cruelty." When we asked her to elaborate on that remark, she began by saying that it is very important that McDonald's has addressed some of the harshest things in the lives of farm animals, like forced molting and how animals are stunned and slaughtered. When it came to factory farming, however, her view was that not much could be gained by tinkering with a fundamentally bad system.

Halverson thought McDonald's should follow the example of Chipotle Mexican Grill, a restaurant chain in which they hold a majority interest. Chipotle is seeking to obtain supplies from alternative farming systems that do not prevent animals from performing their natural behaviors.[8] McDonald's could also look to its own policies in Britain, where there is greater public awareness of animal welfare issues; McDonald's obtains its British eggs from free-ranging hens with access to outdoors. Although supplies of animal products from animal-friendly farms are in short supply in the U.S. at present, Halverson believes that if McDonald's were to provide a market, at a price that made it reasonable for farmers to treat their animals well, the supply would soon grow.

In the long term, McDonald's contribution to funding research at places like Purdue University's Center for Food Animal Well-Being could also make a big difference for animals.[9] Together with Burger King and Wendy's, McDonald's has funded research aimed at scientifically assessing animal welfare and devising farming methods that offer better animal welfare. Other McDonald's animal welfare efforts, however, seem to have stalled. The hens that supply McDonald's eggs still have the ends of their beaks seared off with a hot blade. Back in 1998, Bob Langert told Spira that McDonald's was looking at obtaining its bacon and other pig products from producers who do not use sow crates. But as of this writing, these products still virtually all come from total confinement pig units that keep sows in crates. In Britain, where sow crates are illegal, McDonald's has no difficulty in obtaining its supplies from producers who do not use these methods. Lower standards in the U.S. than in some other countries in which McDonald's operates seem to be part of a pattern. In Europe, McDonald's has quality assurance standards for meat, dairy, and eggs that reaches back to the farm level. In the U.S., the meager standards are limited to eggs.

## A HEALTHIER AND MORE ENVIRONMENTALLY FRIENDLY BIG MAC?

Antibiotics policy is another area in which McDonald's has been a leader. At a time when some agribusiness interests were refusing to accept the strong scientific evidence that the routine feeding of antibiotics to animals could contribute to the development of antibiotic-resistant bacteria, McDonald's issued a Global Policy on Antibiotic Use in Food Animals that was driven by science rather than agribusiness interests. It prohibited the use, for growth promotion in animals, of antibiotics similar to those used in human medicine.[10] Two years after that policy was announced, the U.S. Food and Drug Administration banned the use of the antibiotic Baytril in poultry because of concerns that it could lead to antibiotic-resistant infections in people.[11] As far back as 1989, McDonald's was also ahead of other hamburger chains in developing a policy of not purchasing beef raised on land from which rainforest was recently cleared.

Beyond these points, McDonald's environmental record becomes less clear. McDonald's issues a Corporate Responsibility Report, available on its Web site that, in addition to dealing with animal welfare, also addresses questions of social and environmental responsibility. The 2004 report, the most current as of this writing, strikes all the right notes for a responsible corporation. On issues like employment of minorities and women, the report details the percentage of corporate officers and the percentage of holders of McDonald's franchises who are in those categories. That makes a sharp contrast with the section of the report on environmental policies, which is often strong on rhetoric but lacking in specifics.[12]

The report mentions, for example, that McDonald's Europe has set standards for beef, chicken, dairy, potatoes, wheat, and lettuce, but it does not indicate what these standards are, nor why similar standards have not been developed for the United States and other countries. The report highlights McDonald's ambitious environmental policies, addressing water and air pollution, energy use and greenhouse emissions, soil health, pesticides, and biodiversity, but the language is generally vague and lacking in examples of how the implementation of these polices has made a concrete difference anywhere. On fish, the report states that McDonald's has environmental guidelines on fish sourcing to ensure that the fish it buys come from sustainably managed fisheries. A quote from one of McDonald's European suppliers refers to standards set by the Marine Stewardship Council, but it is not made clear if, outside Europe, McDonald's limits its supply of fish to fisheries approved by the Marine Stewardship Council. Bob Langert, McDonald's director of environmental affairs, declined to answer our questions on these matters.

On these environmental issues, then, it is not possible to assess just how well McDonald's actions match the fine rhetoric of its corporate responsibility reports. In any case, there is an environmental problem with McDonald's business that is far more difficult for it to confront, because it goes to the heart of its menu. If McDonald's is serious about promoting a more environmentally sustainable diet, it should be seeking to move away from corn-fed meat, especially from beef that comes from modern feedlots, which, as Michael Pollan put it, are "floating on a sea of oil."

There is no sign of McDonald's doing that. But it does offer a McVeggie Burger throughout Canada and in some major American cities. (Burger King sells veggie burgers in all its restaurants.) Erica Frank, an associate professor in the Department of Family and Preventive Medicine at the Emory University School of Medicine, has compared the McVeggie Burger with the standard McDonald's burger and says that consumers who switched to the McVeggie would reduce their risk of suffering from diabetes, hypertension, high cholesterol, cancer, and cardiovascular disease. She also regards the soy-based burger as more environmentally friendly. "Besides that," she adds, "both burgers taste pretty similar. So, if you want to pick an easy way to improve your health and the health of the planet, this is a simple and good place to start."[13]

## IS SMALL BEAUTIFUL?

Big chain restaurants have more power than small independent restaurants to advertise, to shape our views about what is or is not good to eat, and to standardize our food supply. They are inherently likely to use that power to increase their profits. But they cannot control what their customers think, and protecting their market share may require them to be responsive to their customers' concerns.

There is much to be said against big multinational corporations, but there is also something to be said for corporations with nationally or globally recognized logos. It wasn't easy for Henry Spira or People for the Ethical Treatment of Animals to persuade McDonald's to improve some aspects of its food supply, but it would have been harder still to run a campaign against 30,000 independently owned and operated hamburger restaurants.

If you are going to eat hamburgers and know of a hamburger joint that buys meat from organically-raised, free-ranging, grass-fed animals, you should go there. But in the absence of that information, a hamburger from a locally owned restaurant is not likely to be any better for the environment, nor for the animals, than the meat McDonald's serves. It could be worse, because the beef could have come from a slaughterhouse not subject to audits like those McDonald's requires; the chicken could have been fed antibiotics that McDonald's does not permit; and the eggs could come from hens with less space than the hens whose eggs McDonald's uses.

## WAL-MART: EVERYDAY LOW PRICES—AT WHAT COST?

If Jake's chicken is an iconic American food, the place she buys it is equally characteristic of America today. Wal-Mart is the biggest everything—world's largest grocer, world's largest retailer, world's largest corporation. Short of being a nation, nothing gets bigger than that. (If Wal-Mart were a nation, it would have a bigger economy than 80 percent of the world's countries.) Each week, 138 million people go to one of Wal-Mart's 5,000 stores in the United States and nine other countries, giving the corporation annual sales of more than $300 billion. With a global workforce of 1.6 million, it is the largest private employer in the United States, as well as in Mexico and Canada. The ethics of what we eat encompasses not only how our food is produced, but also how our food is sold. If so many people do it, can there be anything wrong with shopping at Wal-Mart?

One reason for concern about Wal-Mart is simply its size. Wal-Mart already has 11 percent of all U.S. grocery store sales, and according to Merrill Lynch analyst Daniel Berry, by 2013 that figure is likely to rise to 21 percent.[14] Being the biggest buyer of food gives Wal-Mart a lot of clout over how the food it buys is produced. Do we really want a single corporation to have so great a sway over that? The answer might depend on how the corporation behaves.

No corporation as big as Wal-Mart can avoid criticism, and Wal-Mart gets so much flak that it has set up a "war room" to fight back.[15] So let's focus our discussion on something that Wal-Mart does not deny, and indeed boasts of: the way it seeks to drive down costs in order to provide "everyday low prices." As Wal-Mart CEO Lee Scott said to CNBC's David Faber when discussing Wal-Mart's approach to its suppliers, "The idea is that we say—we sit down with you and say, 'How do we take cost out of doing business with you.'"[16] As a result, according to a UBS-Warburg study, Wal-Mart has grocery prices 17 to 20 percent lower than other supermarkets. Our question is: In constantly striving to reduce costs, has Wal-Mart breached any ethical limits?

Low prices are a good thing. If customers like Jake pay less for their food at Wal-Mart than they would at another store, they have more money to spend on meeting their other needs, or, if they are so inclined, to increase their contributions to good causes. Low food prices are particularly good for the poor, who spend a higher proportion of their income on food than the rich. People living in poor areas often have few places to buy fresh food, and the stores they do have charge higher prices than stores in more affluent neighborhoods. When a Wal-Mart moves in—and Wal-Mart's stores are disproportionately located in the poorer parts of the country—that changes, and the poor can save significant sums.[17]

The positive value of a store with low prices can, however, turn negative if the low prices are achieved by passing costs onto others. In 2004, Wal-Mart's spokesperson Mona Williams told *Forbes* that a full-time store assistant takes home around $18,000 annually. Some think this estimate is generous, but assuming that it is accurate, it still means that if the employee is the only income earner in a family of four, the family is living below the poverty line. According to documents released as part of a gender-discrimination suit against Wal-Mart, researchers found that the average non-salaried Wal-Mart associate in California gets nearly $2,000 in public welfare benefits each year, including health care, food stamps, and subsidized housing. If all California's retailers lowered their wages and benefits to Wal-Mart's level, that would pass an additional burden of $400 million to the state.[18] In 2005 Wal-Mart acknowledged that nearly half of the children of its employees either have no health insurance or are on Medicaid. Wal-Mart itself admits that for the national labor force as a whole, that figure is only one-third. M. Susan Chambers, a senior Wal-Mart executive who led the investigation that produced this finding, admits that she was "startled" by the discovery. Wal-Mart's critics would not have been. Wal-Mart subsequently announced that it would improve the health care benefits it offers its workers.[19]

Nevertheless, Wal-Mart says "it doesn't make sense to say that we cost taxpayers money" and then cites the substantial amounts of federal, state, and local taxes the corporation pays, including sales taxes.[20] But that's no answer to the charge. If Wal-Mart were replaced by stores that paid better wages and gave better health benefits, consumers would still buy food. So the various taxes Wal-Mart now pays would be paid by the corporations or family-owned businesses that sold the food Wal-Mart now sells—and their better-paid workers, instead of needing assistance from taxpayers, would pay taxes themselves.

Wal-Mart's impact on wages and benefits was most clearly shown in 2002 when it announced plans to open 40 Supercenters in California over the next three years. California was then a stronghold for Safeway, Albertsons, and Kroger, three of the largest grocery chains in the country. In contrast to Wal-Mart, which has succeeded in keeping unions out of its U.S. stores, these three chains have unionized workforces, and their workers were paid about 50 percent more than workers at Wal-Mart and had much better health insurance programs. The big three chains believed, with some justification, that these costs would be a fatal handicap in the coming battle to defend their market share against Wal-Mart. They therefore demanded that their workers accept a new contract that contained no wage increases and would require workers to contribute substantially to their health insurance. In effect, the workers were being asked to take a pay cut. The result was the biggest strike in California's history, but against their employers' fear of Wal-Mart, even 70,000 united workers could gain little. After 20 weeks on strike, they agreed to take a contract with reduced benefits. Wal-Mart, or the threat of it, was forcing wages and benefits down.

Wal-Mart applies a similar strategy of cost-reduction to its suppliers. Gib Carey is a partner with Bain & Co., a management consultant firm that has Wal-Mart suppliers among its clients. For many suppliers, Carey says, maintaining their business with Wal-Mart becomes indispensable, but it isn't easy to do so. "Year after year," Carey says, "for any product that is the same as what you sold them last year, Wal-Mart will say, 'Here's the price you gave me last year. Here's what I can get a competitor's product for. Here's what I can get a private-label version for. I want to see a better value that I can bring to my shopper this year. Or else I'm going to use that shelf space differently.'" As business writer Charles Fishman sums it up, "The Wal-Mart squeeze means vendors have to be as relentless and as microscopic as Wal-Mart is at managing their own costs. They need, in fact, to turn themselves into shadow versions of Wal-Mart itself."[21] As a result, suppliers of American-made goods often have to look for cheaper goods made in countries with lower wage costs. If they don't, Wal-Mart will go to a Chinese manufacturer and cut them out entirely. Wal-Mart acknowledges that it bought $18 billion worth of goods from China last year.

Is it wrong to buy goods from China if they can be made more cheaply there than in the U.S.? We don't think so. People in China, Bangladesh, and Indonesia need work too, and since they are, in general, poorer than Americans, they probably need it even more than Americans do. So we are not going to chastise Wal-Mart for buying goods made abroad. The issue is, rather, whether in countries with deep poverty and endemic corruption, the unremitting drive to reduce costs can stop short of sweatshop conditions that include hazards to workers' health, child labor, and debt bondage that verges on forced labor.

In this area, Wal-Mart acknowledges past mistakes. A 1993 *Dateline* exposé showed that clothes sold in Wal-Mart stores under a "Made in the USA" banner were actually made in Bangladesh, and, worse still, were made by child labor. Four years later it emerged that a Wal-Mart line of clothing brand-named Kathie Lee, after ABC morning show cohost Kathie Lee Gifford, was also made by factories that used child labor. Now, however, Wal-Mart insists that it has strict supplier standards that absolutely forbid the employment of anyone younger than 14. Yet Wal-Mart's standards still allow workers to be pushed very hard. Employees may be made to work for 72 hours in a six-day week—that's 12 hours a day, for six days solid—and to work for 14 hours in a single day.[22]

In the minds of Wal-Mart's critics, however, the problem is more one of enforcement than of the standards Wal-Mart espouses. Some corporations, like Gap, allow independent organizations to inspect the foreign factories of its suppliers. Wal-Mart prefers to use only inspectors it hires. Their impartiality can be questioned.

Wal-Mart is a member of the Food Marketing Institute and as such has been taking part in the discussions that this Institute and the National Council of Chain Restaurants have been holding to set standards for animal welfare. As we saw when looking at McDonald's, the move to set standards was triggered by measures taken first by McDonald's and then matched by other chain restaurants. In many areas of animal production, the FMI/NCCR standards do little more than describe existing factory farm practices, although in a few places they tinker with some of the details, to the modest benefit of the animals. For example, they adopt wholesale most of the National Chicken Council's Animal Welfare Guidelines, which, as we have seen, give chickens a space allowance the size of a standard sheet of typing paper, permit their beaks to be seared off, and allow the breeder birds to be kept half-starved. They also allow catchers to pick up the birds by one leg and dangle five live chickens from each hand.[23] Regrettably, as we have already seen, there is a tension between strong animal welfare policies and "everyday low prices." We'll learn more about that in Part II, when we consider the animal welfare standards being set by Whole Foods Market, a food retailer at the other end of the spectrum from Wal-Mart.

In his book *The United States of Wal-Mart,* John Dicker writes that the success of Wal-Mart says something about us: "The cult of low prices has become so ingrained in the consumer culture that deep discounts are no longer novelties. They are entitlements."[24] Bargain-seeking seems to be such a basic aspect of human nature that to question it can appear quixotic. But at Wal-Mart, the bargains hide costs to taxpayers, the community, animals, and the environment. That is why, despite the undoubted benefits of Wal-Mart's low prices, a very large ethical question-mark hangs over buying our food at Wal-Mart. With a global workforce of 1.6 million, it is the largest private employer in the United States, as well as in Mexico and Canada. In Britain, ASDA has been wholly owned by Wal-Mart since 1999. In 2003, ASDA overtook Sainsbury's as the second largest supermarket chain in Britain, behind Tesco.

# PART II

# THE CONSCIENTIOUS OMNIVORES

# 6

# JIM AND MARY ANN

When Roger Ludlowe, one of the founders of the colony of Connecticut, arrived in 1639, he found already-cleared fields abandoned by native Americans who had nearly been wiped out by diseases introduced by earlier European explorers. The settlement he established soon prospered, its name acknowledging the bounty of its lush grasslands: Fairfield. A century of busy trade through the town's fine harbor brought wealth and a spirit of independence, for which Fairfielders paid a high price during the Revolutionary War—in 1779, 2,000 British troops burned it to the ground.

Today's Fairfield straddles a noisy tangle of expressways, parkways, and rail lines linking New York and Boston. Among the town's 57,340 residents is the family of Jim Motavalli, his wife Mary Ann Masarech, and their daughters Maya, aged nine, and Delia, seven. We chose them because of the ethical principles—leaning to green, coupled with concern for animal welfare—that guide their food choices. Mary Ann and the girls eat meat and animal products from humane and organic farms. When we asked her whether she developed good eating habits while growing up, she laughs and says "We ate Spam! I absolutely loved grilled Spam. My mother made what she called "red spaghetti" but it looked orange—it was Campbell's tomato soup and milk and Velveeta cheese mixed together over spaghetti. And that would go with the Spam." Both her parents worked, she explained, and her mother came home from work at the same time as her father. That meal was something she could get on the table fast.

Jim, an environmental editor and writer, avoids meat because he believes that the prevailing practices in the meat and poultry industries are "incredibly cruel and wasteful." He comes from a very different background from Mary Ann. When Jim was 11, his family moved to Iran. While at high school, he spent two years in Iran, where his father, who was of Iranian descent, worked for the U.S. Agency for International Development. During the two years the family spent there, he never really took to Iranian cooking, except for the rice. The family then returned briefly to America before shifting again for another two years, this time to northern India. That was when Jim was first exposed to vegetarianism, since a large proportion of the population of India is vegetarian. It was not until some years later, however, that he stopped eating meat. That decision was based more on his environmental concerns than on animal welfare grounds.

Fairfield is a bead in the string of affluent towns like Westport, Darien, and Greenwich that lie along the Connecticut coast toward New York City. Many Fairfielders are commuters who spend up to three hours a day on MetroNorth trains to maintain high-end jobs in Manhattan, sixty miles away. Jim and Mary Ann's relatively modest two-story colonial-style house is in the Brooklawn area—a down-to-earth, middle-class section where Fairfield borders Bridgeport, a city of decayed industries that actor Paul Newman once called "the armpit of Connecticut." Around the small in-ground pool in their backyard are gardens of tomatoes, peppers, horseradish, raspberries, blackberries, blueberries, and herbs like rosemary, basil, cilantro, and thyme. "All organic," Mary Ann says as we tour the yard with Delia and Maya, who keep showing us their favorite plants and foods.

Mary Ann is slight and wiry, with shortish hair and intense dark-brown eyes. She's a product manager with BlessingWhite, a corporate training and consulting firm with the mission of "reinventing leadership and the meaning of work." She mostly works from home but travels regularly to the firm's headquarters in Princeton, New Jersey, as well as to visit clients around the U.S. That means she no longer has time to do as much cooking as when she and Jim first got together, almost 20 years ago. As we walk through the house and grounds, she moves about almost as one with her daughters, chatting with them all the time. They're brunette cherubs with pigtails and round, cheerful faces. We've all just come back from a shopping trip at Trader Joe's, one of the nationwide chain's 200 "specialty grocery stores." According to its Web site, "All our private label products have their own 'angle,' i.e., vegetarian, Kosher, organic or just plain decadent, and all have minimally processed ingredients." But in terms of permission to record our shopping trip, whether by video camera or tape recorder, Trader Joe's proves no different from Wal-Mart in Mabelvale, Arkansas: the Fairfield store manager's response is a firm "no."

We unpack the groceries and lay them on the table. Mary Ann takes each package and reads from the label: "This is Genoa salami from Applegate Farms. It says it's uncured, no antibiotics. It's $2.99, so it's not cheap. 'American Humane Association monitored'. It says it's from farms that allow animals. . . . Let's see: 'adequate shelter, sufficient space, the ability to express normal behavior,' So we're eating salami from pigs that have got a chance to express who they truly are!" Then she picks up a pack of bacon. "Niman Ranch Dry Cured Center Cut Bacon. Just listen to this." She reads from the side of the box: "'Niman Ranch produces the finest-tasting meat in the world by adhering to a strict code of husbandry principles. Our heirloom pork comes from hogs raised on pastures or deeply bedded pens. Our small family farm raised hogs are humanely treated, fed the finest feeds, never given growth stimulants or antibiotics, and raised on land cared for as a sustainable resource.' It's pricey, at $4.99 for a 12 ounce pack, but it also says that it will be the best bacon you've ever tasted, and Maya and I do like it, it's nice and thick."

Now she takes out a small carton of eggs. "Here's our favorite—Pete and Gerry's organic eggs." The price tag is $1.59—for a half dozen. The label says they come from cage-free hens in New Hampshire, raised without medications and antibiotics, and fed a diet of organic food grown without pesticides. "It's got the USDA organic seal. And it's 'Certified Humane.' This is one of the few products that gives you all those features in one."

She picks up a half-gallon of milk and reads from the label: "'Trader Joe's organic milks are produced with milk from local family farmers. The milk is certified organic and is governed by rigid standards in every aspect of organic milk production. Trader Joe's organic milks come from cows fed a strict organic diet. Neither the cows nor the feed are ever treated with hormones, antibiotics or pesticides.'

"Here's Soy Delicious, a soy-based ice cream. We eat this because the kids like it. They don't always accept alternatives to dairy and meat. There's another one, Tofutti Cuties, that they like. But anything like fake hot dogs? Jim will eat them but you can't get them by the girls." As a result, the shopping also includes beef hot dogs from Hebrew National.

Mary Ann doesn't always do her shopping at Trader Joe's. In summer she often goes to Sherwood Farms, "a real family farm, hand-painted sign, you purchase produce up back behind the house. You never know what you're going to get there: eggplant, cucumbers, peppers—all kinds of peppers—green beans, cauliflower, corn. I supplement our own tomato crop with their tomatoes. I *love* going there."

The phone rings; Jim has arrived at the train station and Mary Ann has to leave to pick him up. "He's been in Switzerland with a bunch of journalists looking at an energy plant, sustainable factories, and other sorts of environmental things over there."

When they return, Mary Ann, Maya, and Delia start putting dinner on the table while we talk to Jim. He is a rare combination: a syndicated auto columnist and an ecologist. He edits *E*, a bi-monthly magazine on environmental issues, and has put his interests together by writing *Forward Drive,* a book on "clean" cars, and *Breaking Gridlock,* on environmentally friendly transportation. He also edited *Feeling the Heat,* on global warming, and at the time of our visit was editing a fourth book, *Green Living*, a guide to living lightly on the earth. As if all that was not enough, Jim travels frequently to speak at conferences, hosts a weekly radio show on noncommercial WPKN featuring public affairs and folk, jazz, and blues music, and still spends a lot of time with his family. We envied his calm. Just back from two hours of train rides after a seven-hour flight and six time zones of jet lag, he's joshing about the family's ethnic mix.

"My half is half WASP. Old New England. Chadwicks, Biddles, and Barrettes. One great-grandfather was the oldest living graduate of Harvard Medical School and was a pioneer of a tuberculosis cure, Henry Dexter Chadwick. The other side is Iranian. Aristocratic there, too," Jim chuckles. "They can wear the green turban, indicating direct descent from The Prophet." He adds that one grandfather ruled over a city, Sabzewar, in Iran.

"Not to mention having four wives," Mary Ann chimes in. Jim adds that his parents met in Japan and that Mary Ann is Slovak.

"Well, half," she says. "My mother is Irish and German. I provide the peasant stock in the family."

"Good, sturdy peasant stock," Jim says, still chuckling.

Dinner caters to everyone's tastes and principles. It begins, for Jim and the girls, with bowls of soup made with organic butternut squash and carrots. Meanwhile Mary Ann crumbles pieces of organic feta cheese over a big bowl of organic greens. After that, Mary Ann and the girls are having a filet mignon wrapped in bacon. These three describe themselves as "caring carnivores"; that is, they eat meat from animals raised according to humane or organic standards. Jim eats vegetarian but will have a piece of fish on occasion, depending on what he knows about the fishing practices. Tonight he is eating broiled Norwegian salmon served with couscous.

That's not all. Maya jumps up and goes to the stove to tend to a favorite dish she has invented: chick peas with "Motavalli sauce." Jim asks her how she makes it. "Well, you cut up the garlic in little pieces and fry it in olive oil a little while and then you dump in a can of chick peas. You cook it a little until the chick peas are nice and hot, then you add the juice of a lemon, some salt and pepper, stir everything all around . . . and VOILA!" The others laugh and clap applause.

"Very good, Maya," Jim says. "You should have your own cooking show."

Over this impressive gourmet meal we talk about the thicket of ethical concerns that shape the Masarech-Motavalli's meals. "People want ethical choices," Jim says, citing his experience writing about cars. "They want to make the environmental choice, but they're not willing to do any trade-offs for it. They still want it to be as cheap as the other things they could buy." Mary Ann agrees, saying, "If we're talking food, it has to taste as good. I want an ethical choice that tastes the same." She talks about the complications in balancing ethics with taste, convenience, cost, and other considerations. "I work so much that I need that convenience. If I had more time on my hands, I'd probably make better ethical choices, but I would have to go out of my way, which I can't do much of right now."

They tick off some of the other ethical concerns that govern their choices: how well a food producer scores in the areas of labor relations, corporate responsibility, animal welfare, and environment. Then it gets more complicated as we break each of these down into various concerns. Under environment, Jim notes energy efficiency, water use, waste management, toxic by-products, and other concerns. "That's one major thing

I have against meat production; it's just incredibly wasteful and environmentally degrading. It takes eight pounds of grain to produce a pound of meat."

Maya and Delia have stopped eating and are staring at their father.

"I know," says Delia. "But it's just hard not to eat meat."

Maya: "I could . . . for the cow."

Jim: "Delia, why do you say it's hard to . . . ?"

Maya: "I'm not going to eat any. . . . I could give up all the meat except steak."

Jim asks Delia again.

Delia: "I really like bacon and steak."

Maya: "I like fish."

Jim: "You really like the taste of it you mean?"

Delia: "Yeah."

Maya repeats that she could give up anything but steak. Jim asks her what kind of steak she likes: "Rare," she blurts out laughing, and we all laugh with her.

Later, after the girls have gone into the next room, Jim tells me that this is one of his ongoing arguments with them. "I'm constantly telling them: it isn't a matter solely of what tastes good, because that means you end up with lots of sugar and lots of salt in your diet." He pauses, laughs, and shakes his head. "But getting them to agree with that is very difficult." He thinks it's a lot easier to shop ethically if you're not a parent. "Because when you are a parent, what the kids will eat is a major consideration. And what they want to eat is influenced a lot by marketing."

"I'm stuck in the middle," Mary Ann says, "because we kept a vegetarian house for ten years before the kids were born, even though I wasn't a vegetarian. I never cooked meat in the house. Occasionally I'd cook fish. But most of the time I didn't cook any animals at all. It worked fine. I liked to cook. We liked to do different stuff. And then in the last six, maybe nine years, it's just gotten harder because I'm working more, I have more palates to please, and that's where the meat comes into the picture." She wants me to note, though, that they do have several vegetarian meals each week.

"See, here we have a division in the family," says Jim. "Because bacon seems like an unforgivable product to me."

Mary Ann: "Well, bacon's so wrong on so many levels, because ethically. . . . "

"How can you say that and yet you make it for them?" Jim cackles, before answering his own question: "Again, they like the taste."

Mary Ann: "There are a couple of reasons. One is, I send them to school and I don't know what they're going to eat. Half the time it's something they don't like. I want to get something substantial in them."

Jim: "Yeah, like salt and nitrate," laughing openly.

Mary Ann: "I didn't cook any bacon all summer. Then we had these great tomatoes in the garden. All organic. All kinds. When we got this excess of tomatoes, I went out and bought bacon because I wanted a B.L.T. I was dying for one!"

We return to seafood. Jim is very concerned about fishing practices, from an environmental standpoint. He talks of the need for sustainable fishing that doesn't cause fish populations to crash. Mary Ann prefers fish over meat for a different reason: "I would much rather have the kids eat a little fish that isn't going to look me in the eye with any sensitivity rather than a cow. So for me, fish is preferable to a mammal just because . . . it's a fish."

The conversation veers in yet another direction. Jim is concerned about the current trend in which the big food corporations are buying up organic, soy-based, and alternative food companies. "I think it's a bad thing because the bigger the company, the more they are going to look to cut costs and corners and take everything right back to all the usual mass production techniques in order to make more profit. Take organic foods. If you look at the founders of the organic food business, they were more motivated by ethical considerations. But the big, powerful companies who are buying the businesses from them are not as likely to be so motivated. They are going to be much more likely to strain the limits of what constitutes 'organic' production. And they'll lobby hard to relax those standards so that they can cut more costs and make more profit. The bottom line. That's just the law of gravity of big business." There's no sign of good-natured humor in him now. "I think it's irreversible. The company that cuts the most corners is able to put the product on the shelf at the lowest price. So they'll get more market share."

"I know, but you're a purist," says Mary Ann. "I work in the corporate world, I have corporate clients. Generally they won't do things that harm their best interests. The reality is that if enough consumers want an ethical product like we've been talking about, they'll figure out a way to do it. That's a new market for them, and business loves new markets."

From there we shift to international trade, and I ask about the coffee we are drinking. Jim tells me it is something that was sent to the magazine—they are always getting stuff sent in for review, including coffee. "Ideally, coffee should be 'triple certified,'" he says, "that means, organic, fair-traded, and shade grown. Shade grown means that they grow the coffee under shade trees, which is more bird friendly. I think this one has all three certifications. But we also drink a fair amount of supermarket coffee."

"I would buy more from Trader Joe's," Mary Ann adds, "but they sell only coffee beans, and I never feel I have the time or inclination to grind coffee. Convenience again."

It's quiet for a moment, except for Delia and Maya in the next room screaming that they want to get on with the surprise dessert they have planned.

Mary Ann sighs. "Like we've been saying, it's all about trade-offs. As you can see, we go back and forth on this stuff. Jim's much more of a purist than I am. I'm the pragmatist. If we were to live more by our principles, then I would probably not want to live where we live, in one of the most expensive places in the country. I would probably choose not to work or I would work in a way that I could have more time, so I could spend more time growing and getting and preparing food. I believe we could be more environmentally conscious, more humane, more politically correct in everything we do. But we like our lifestyle in this area. It's really the social network of friends and family that keeps us here."

Heartlessly ignoring louder and louder pleas for dessert, we raise questions about how much they, as ethical but practical consumers, trust the claims and the companies behind the foods they eat. We ask Jim and Mary Ann: take Applegate or Horizon or any of the other makers of the food you buy. It advertises that food product as ethical and it charges a premium for it. Is that company open and accountable to its consumers? Will it open its doors to you? What will they do if you say, "I buy your organic milk from happy cows on pasture and I want to meet you and your cows and see them being milked."?

"I would love to be able to do that," Mary Ann—corporate-friendly pragmatist—says, "but knowing the corporate world as I do, they don't want you to take a look under the hood." She believes it is best to buy food locally, because that way you're more likely to have some first-hand knowledge about the producer and its practices.

"And the food's fresher and riper," says Jim, "and hasn't cost an ecological fortune in fossil fuels coming across the continent."

"And it might contain fewer preservatives and the chemicals put in to survive some hellish truck ride," adds Mary Ann.

"It's going to get better," she says. "There's some sort of a circle to it. As more consumers get aware, they make demands and create new markets. Then more companies get interested and are willing to invest in these alternatives. Then consumers have more options available to them. Then more and more corporations invest and get on the bandwagon. Little by little there are more food choices available and affordable. So when Jim rails against corporations, I'm out there going, 'well, that means it will be easier for me to get and it will be cheaper and it will be more widely available to everyone else out there.' If you want to make change, you've got to make it easier for people."

Consumer demand suddenly appears in the form of Maya and Delia, carrying a pot of melted chocolate and a bowl of fresh strawberries. "It's all organic—the chocolate and the strawberries," Mary Ann says. Maya assures us that this is chocolate from Trader Joe's, which imports it from countries that do not use child slave labor in harvesting cocoa beans. But no one seems to be pondering that as they eat—in less than a minute, it's all gone.

There is no well-established label for the kind of diet that Jim and Mary Ann follow. Michael Pollan has suggested the term "humanocarnivore" for someone who eats only humanely-raised meat. But he admits this is not a catchy label, and he is surely right about that. Roger Scruton, a British philosopher, writer, and defender of meat eating, has written an essay entitled "The Conscientious Carnivore," and this comes closer to being an accurate description of the kind of person we have in mind.[1] But "carnivore" is not quite right, because no one is suggesting a diet consisting only of meat. That is why we prefer the term "conscientious omnivore" to describe someone who eats meat or fish, but only when it satisfies certain ethical standards. Just what those standards should be, of course, is something we still need to explore.

# 7

# BEHIND THE LABEL: NIMAN RANCH BACON

We are on Mike and Suzanne Jones' 73-acre pig farm near Louisburg, North Carolina, hunkered down in the shade of some trees, watching a sow doing something that would amaze anyone who knew about sows only from factory farms. This sow lives with several others in a two-acre enclosure that includes an open field in pasture and woods along the southwest side. A half dozen little arched huts known as "English arcs" are scattered along the edge of the woods. This particular sow is ambling about in the brush along the tree line, biting off twigs and leaves and carrying them back deeper into the woods where she is building a nest. It looks like a large bird's nest on the ground, about five or six feet in diameter, a ring of tangled small branches, leaves, and detritus from the forest floor. The sow carries a leafy branch into it and tucks it in among the others. She goes back, snaps off another branch, and this time carries it out to one of the huts, as if she has changed her mind. Mike says they do that sometimes. "The sow that had her pigs yesterday made two different nests and then decided which one she liked better. This one may do this too. The other sow that had her pigs this morning made her nest under the honeysuckle vines."

Mike left the world of corporate confinement pig production to produce his own pigs on pasture and to work for North Carolina Agricultural and Technological State University in a program to help small farmers establish low-cost pig operations like his own. Also with us are the Halverson sisters, Diane and Marlene, who advise the Animal Welfare Institute on its Animal-Friendly Husbandry Program and also serve as consultants to Niman Ranch. Mike's farm raises pigs for the Niman Ranch system, and the Halversons have invited us to come along on one of their periodic inspections to ensure that the farms are conforming to AWI's strict standards for animal-friendly farming. The Halversons live on their family's farm in Minnesota and aren't at all surprised by what we are seeing. Sows about to give birth, Diane tells us, have an instinctive drive to build nests from materials like leaves or straw. The nest will provide a comfortable place for her to snuggle into when she gives birth to and then feeds her piglets. In fact, allowing sows the ability to build nests—having both space and the freedom of movement to build the nest from materials natural to the environment or provided by the farmer—is a requirement of the AWI standards, because it is regarded as a key element of good sow welfare.

The sow does not appear to be in a hurry, so as we sit and watch, we chat about Mike's former life working in the confinement buildings of one of the nation's largest vertically integrated corporate pig producers. Mike tells us how he had to move pregnant sows from the crates of the gestation buildings to the farrowing crates in another building. "There were these long, slippery concrete runways. The sows were late in gestation and they weren't very comfortable. They weren't in very good shape because in their stalls they hadn't been able to move around. It was very stressful for them." Mike would use a high-pressure washer to clean off all the feces that had accumulated on the pigs while they were in the gestation stalls and then disinfect the sows before putting them in the farrowing crate. Mike says that a sow able to build a nest is less anxious than one confined to a crate, who has no way of relieving the drive to nest. In his experience, sows able to move around when pregnant are also stronger and healthier than those kept immobile in stalls.

Talk turns to the various animal welfare standards that are being advanced as consumers become more aware of, and critical of, mainstream pork industry practices. Mike tries to follow the AWI standards in his work of helping farmers set up pasture pork farms. "I encourage them to conform to AWI standards, because if they do, they should be able to comply with any other kind of niche market opportunity that may come up for them." We talk about consumer confusion over labels and claims that arise as more and more sets of welfare standards are advanced. Besides AWI's program, there is Certified Humane, which we discuss more fully in a later chapter. Then there are the programs behind the label claims of Organic Valley, Applegate Farms, Whole Food Markets and other food companies. A mention of the National Pork Board's Swine Welfare Assurance Program drew a sharp response from Diane Halverson: "There is no such thing as 'welfare assurance' in a program that sanctions crating pregnant sows so they can't walk, or forcing pigs to live 24/7 on bare concrete floors with only bars and blank walls for vistas!" There are many scientific studies documenting the damage done to animals by these housing methods, she told us, adding: "To call a set of standards that permits these conditions a 'welfare assurance program' is a perversion of the truth."

We walk through some of the other pig pens on Jones' farm. Mike shows us a small, uprooted elm sprout that has been stripped of its bark. The pigs, he says, love elm bark for some reason, so he gives it to them for a treat. We're in one of the pastures where the pigs are "finished" to market weight. There's not much grass or clover here. The ground is mostly bare and pocked with mud holes that have been rooted out by the pigs. Some farmers with pigs on pasture put rings through their noses because that causes discomfort when the pigs root around and so reduces damage to the pasture, but Mike doesn't do that. "I don't care if they devastate it," he says. "I've got that tractor and I can fix everything up and plant it in grass again." He laughs. "I don't see why we spend so much time and money trying to fight nature. Seems crazy to me. If they want to root, they are going to root—even if that ring's in there. They will just rip it out, and it will hurt." It's the same philosophy he follows with his sows and nursing pigs. He leaves a litter with their mother to nurse for six to nine weeks, as opposed to a couple of weeks in intensive pig systems. "She's going to wean them naturally at nine weeks anyway. It saves me money if I don't have to buy expensive starter rations for weaned pigs." Someone makes a remark about how he does not spend money to control nature. "Right," he says. "I adjust my management around what pigs do. Rather than try to make them fit into my box, I build my box around them."

A few weeks later, mindful of the restrictive farrowing crates we saw on Wayne Bradley's intensive pig farm in Iowa and the argument that these crates are necessary to prevent the sow smothering her piglets, we ask how many piglets the sow had and how many were smothered. The answers are fourteen and none.

## PIGS IN CLOVER

The Holmes family—Tim, Mike, and their father, George—also produce pork for Niman Ranch, on 360 acres near Albemarle Sound on the North Carolina coast. Tim starts the tour by showing us his pregnant sows—some grazing, some sleeping in a lush, green field scattered with small A-frame huts made of weathered plywood. He points to a barren, empty field beyond them and explains how he moves the sows from pen to pen, depending on weather and pasture condition. "They ate that down, and we moved them in here. They love this clover." We all go quiet for a moment, just watching the sows' activity. "This is about a nine-acre field and there's roughly forty sows in here," Tim says, and then adds with a chuckle, "That's a little more space than a crate!"[1] At the Holmes' stocking density, each sow has almost 10,000 square feet of space. That is a square about a hundred feet on a side. That single sow has as much space as a typical factory allocates to 700 sows confined in gestation crates.[2]

The Holmes have 120 breeding sows and sell about 2,000 market pigs a year. All of them live outdoors in pens. The fencing is electric: two wires, one about knee high and a lower one about six inches above the ground, powered by a small solar panel rigged up to a 12-volt battery. It's cheap and easy to move from pasture to pasture. As for the pigs getting electric shocks, Tim points out that they aren't going near the fence: "Once they learn, they've learned." He points out a shallow trench that runs along the sides of the pens to catch any flow of wastes during heavy rains. There is a mild smell of pigs and manure, but it does not have that heavy, suffocating odor that emanated from factory pig farms we drove by in Iowa. Tim motions for us to look down in the grass, where he points out some pig dung. It is heavily fibered and dry from exposure to the sun. He explains that the chlorophyll in the pigs' diet from grazing produces "pasture poop" that is not as smelly as that from grain-fed factory pigs. Almost all of the sows had their heads down, busily grazing the fescue grass and clover. Diane points to one, saying we can see that pigs are grazing animals as well as rooting animals. She tells us about studies of domestic pigs allowed to live in a forested enclosure near Edinburgh. The researchers found that during daylight hours, the pigs spend most of their time either grazing (31 percent), rooting (21 percent), or examining and working over the area, exploring and manipulating objects they found (23 percent). That comes to three-quarters of their day.[3] The idea that pigs like to lie around all day is wrong—they only become "lazy pigs" when we confine them so that they have nothing to do.

Tim also feeds them grain every day; he scatters it right on the ground all around the pens. The idea, Diane explains, is to make sure that each sow gets some feed. "They have to work at it and get through the grasses to get at the feed. That way the boss sows don't 'hog' it all." That, she comments, is one of the main excuses that industrialists use for putting sows in crates—to keep the more aggressive sows from fighting the others away from the feed. "But this is the sane solution: let them have enough space and spread the feed around." One sow comes trotting across toward us. "You can see how well they can walk and trot, with no lameness or soreness from hard floors."[4]

Diane elaborates on the principles behind the AWI standards for these pasture pig farms. "The basic one is that the pigs have to be able to fulfill essential elements of their normal behavior. That is determined, really, by both practical farm experience and the research that's been done on the behavior of domestic pigs in the kind of environments that their wild ancestors came from." There are two key elements for sows, she says. The first is to live in a social group, and the second, as we saw on Mike Jones' farm, is to build a nest.

We ask Tim about the labor involved in his pasture operation. "It's a lot of feeding, a lot of sorting pigs. And with this system, it's a lot of fence running. It's putting fence up, getting fence down. It's moving shelters and moving animals. We tell him that we see tails on all of the pigs. "That's not an issue out here," he says. "There's really no reason to dock tails if you're keeping your pigs outside. We have had them inside in the past, though, and I can say that you will run into some trouble inside." We remark that pigs indoors are smart animals confined in a sterile environment. "And that's it, too," he says. "The standard used to be eight square feet of space per pig. A hog's a curious animal, so they'll look for something to do and. . . ."

Our next stop is the farrowing area, another field of about two acres, enclosed by an electric fence. The Holmes' farm uses the same "English arcs" as Mike Jones. "We try to bring the sows in here at least a week before they are going to farrow," Tim says. "We like to put about eight to a pen, maybe as many as ten." We look at one of the huts up close. It is about eight feet square at the ground and chest high at the top of the arch. It sits right down on the ground; there is no floor, just a thick bedding of hay. "You don't want that floor," Tim says. "You want that hay. And then when we move it, the sunshine will kill any kind of pathogen that's in there. Sunshine is a beautiful thing. And that's something that you can't use in confinement. I can tell you, when we were raising them inside, every group of pigs we took in there, the first thing they'd do within a week was scour." ("Scour" is the term farmers use for diarrhea.) Every group. And we had to break out the neomycin or whatever antibiotic we were using. We had to treat every group. But in an outdoor setting like this, I never had any trouble."

The piglets remain with their mothers in the farrowing area for about six weeks before they are moved to other pens for feeding to market weight. The farrowing pens are alive with their youthful energy and constant motion. Some flock around a sow, annoying her with their demands for milk; others join up with other litters of the same age in racing up, down, and all around the pen.

We bring up the subject of castration of the male baby pigs. Tim says it is the market. "Hey, if I didn't have to cut them, that would be just wonderful." But this is one issue on which the organic farmers agree with Wayne Bradley. U.S. consumers won't accept meat from intact male pigs, so there is no alternative.

## THE MAN BEHIND NIMAN RANCH

The day before we visited the Holmes family, we'd met Bill Niman and the Halversons in Chapel Hill. Bill had regaled us with the tale of the unconventional origins of Niman Ranch—a business that now has annual revenues of about $50 million. Bill is from Minneapolis and graduated with a degree in anthropology from the University of Minneapolis. He went to Berkeley to do graduate work at the University of California. It was the 1960s—the days of the Vietnam War, but also of flower power and hippy culture—and Berkeley and San Francisco were at the heart of it. Instead of going to war, Bill was able to do alternative service teaching school in one of California's agricultural districts, so he got to know something about agriculture. Later he became part of a group that wanted to set up an alternative kind of society, growing their own food and living in harmony with nature and with each other. Bill and his wife moved up to Bolinas, about 25 miles north of San Francisco, where, living in a kind of commune with other like-minded people, they kept chickens, goats, pigs, and horses. Orville Schell was their neighbor—he's now dean of the school of journalism at Berkeley—and he was raising pigs on about 11 acres of land. Niman became his partner, and they sold sides of pork to neighbors and the local community. Bill's wife started tutoring the children of some of the local ranchers, who raised Angus and Hereford cattle. When they had a couple of surplus calves, they gave them to Bill and his wife. They were traditional ranchers who took care of the land and didn't use antibiotics or hormones, and Bill and his wife learned from them.

Bill's wife was killed in an accident, but Bill and Orville continued raising pigs and cattle. At one point, Bill says, they wanted to be like "modern" pig farmers. "We thought, 'We're going to expand this operation and raise a lot of pigs.' We traveled to Iowa. We were pretty seduced . . . we wanted to be like those guys. We built this barn with flush gutters and gestation pens. . . . Then realized, 'Wait a minute, this isn't right' and we never put any pigs in the building. We ended up raising shitake mushrooms in it." Niman and Schell kept raising pigs outdoors, as well as cattle and lambs. They started selling to Chez Panisse, Zuni, and other upscale restaurants. Word got around and the business expanded, but Schell, who had studied Chinese history at Berkeley, was also pursuing a writing career. (He has published 14 books, including several influential volumes on China, and an excellent, if now dated, book entitled *Modern Meat: Antibiotics, Hormones and the Pharmaceutical Farm.*) Eventually Niman bought him out.

As the demand for outdoor-raised pork outgrew his capacity to supply it, Niman made contact with Paul Willis in Thornton, Iowa, who was raising pigs on pasture. "We tasted his pork," Niman said, "and it was exponentially better than what we were doing in California." Around 1994, Niman began selling Willis's pigs one week and their own California pigs every other week, but they moved over to selling only Willis's pigs. When the demand exceeded his supply, Paul began looking for other pig farmers using methods similar to his own. He had heard that the Halverson sisters had, together with the Animal Welfare Institute, developed a set of standards for the humane rearing of pigs. Willis contacted Diane and Marlene, and they helped him find other farmers who met those standards. In that way, Niman Ranch added one farm at a time, until now they are supplied by 470 active pig farms in 15 states, providing them with 3500 pigs a week.

Niman Ranch pays its farmers 5 cents per pound above the going market price—for example, in July 2005 the market price was 47 cents per pound, so Niman was paying 52 cents. Niman also pays an additional bonus that varies according to the quality of the meat and some other allowances. The handling and processing costs associated with Niman meat are higher than the industry average, too, partly due to Niman's relatively low volume. The result is that by the time it gets to the stores, Niman Ranch bacon can cost twice as much as factory farm bacon. The Oscar Mayer bacon Jake bought cost $4.59 a pound, while MaryAnn's Niman Ranch bacon was $6.65 a pound at Trader Joe's, and can cost even more in some stores. That price difference reminded us of our conversation with Wayne, the Iowa intensive pig farmer. When we told him that we were looking at open range pork production, too, and wanted to speak to someone who had a different point of view, he was characteristically direct: "Well, I definitely have a different point of view on that. Whether it's right or not, my philosophy is that there has to be a place for the average citizen who's making the average wage to be able to go and buy a good healthy product. I take a lot of pride in the fact that our product doesn't demand ten dollars a pound for hams like Niman Ranch gets."

We told Bill and the Halversons the gist of what Wayne had said. Bill responded: "It's this whole American thing about having cheap food. It's a fallacy. That guy thinks his food is cheap, but you and I are subsidizing that cheap food by paying for the social and ecological issues that are occurring in that community." "Not to mention the animals," Diane added. "The farmer should always get a decent reward for having given decent treatment to the animals. They need that in order to stay interested in raising animals well and to keep another generation on the farm. Meat and the lives of animals should be treated with much more regard. People should be prepared to pay more and not think of meat as an everyday, throw-away food."

The pigs at Mike Jones' and Tim Holmes' farms are living in a different world from that inhabited by the pigs in Wayne Bradley's sheds—even though, among intensive pig producers, Bradley must be one of the best. Unlike almost all the big producers, he avoids the use of sow crates, and because he is relatively small for an intensive producer, he keeps more of an eye on what is happening than the managers of many larger units. Nevertheless, his pigs, confined permanently indoors on bare concrete, cannot, as Mary Ann put it, express who they truly are. On the farms that supply Niman Ranch, the pigs really can be pigs, and they seemed to be enjoying that.

Right now, most pigs in America can't really be pigs. As we saw in Chapter 4, more than 90 percent of them are in total confinement, never getting to walk around in a field. Economics and the consumer demand for cheap food continue to drive the pork industry. Are ethics and affordability in irreconcilable conflict here? We'll return to that issue in our final chapter, after considering some other possibilities.

# 8

# BEHIND THE LABEL: "ORGANIC" AND "CERTIFIED HUMANE" EGGS

Pete and Gerry's organic eggs are Mary Ann's favorite. The carton proclaims that they are laid by cage-free New Hampshire hens, fed on a vegetarian diet with no medications, antibiotics, or pesticides. In addition to the USDA "organic" seal, the carton has a "Certified Humane" logo, asserting that the eggs have been produced in accordance with standards for humane farm animal treatment. When we wrote to Pete and Gerry's Eggs, Jesse Laflamme responded promptly, saying we could visit whenever it suited us. It was early October when we got there, a sunny but brisk morning. We drove through picturesque New England villages and hills ablaze with the colors of fall, scarlet and orange sugar maples and yellow birches, until we reached the lush green valley in which Pete and Gerry's farm was situated. We couldn't see any hens enjoying the grass, though, just several large sheds that took up most of the farm.

Jesse was expecting us. Ginger-haired, with metal-rimmed glasses, wearing a gray sweatshirt with "Pete and Gerry's Organic Eggs" around a logo picturing green hills, he was younger than we had expected. He told us we were welcome to see, and photograph, whatever we wanted and to ask about anything at all. He did ask us if we had been on a poultry farm recently, but when we assured him that we had not, he said that we could walk right in among the hens—otherwise he would have asked us to observe them through a window.

We had seen some organic egg farms in which hens range freely in well-grassed meadows. We knew that Pete and Gerry's hens were not kept like that—the eggs were not described as "free range." Nevertheless, when Jesse took us inside the nearest shed, our first glimpse of the hens was a shock. The shed was about 60 feet wide and 400 feet long. Covering the floor, stretching away into the distance, was a sea of brown hens, so crowded that the shed floor was visible only down the center of the shed where the hens had left a gap in between the feed and water areas on each side. There were, Jesse told us, about 20,000 birds in that shed. Each of them had 1.2 square feet, or 173 square inches, of space. It seemed very little, though it is more than twice the space recommended by United Egg Producers for their "United Egg Producers certified" program and more than three times the space some U.S. egg producers allow their caged hens. More important still, these birds were not in cages and could actually walk around a large area and flutter up to perch on the water pipes and feed hoppers, even if they had to press through other hens to do it.

We walked in among the hens, trying to look at them as individuals rather than just a vast flock. Battery birds often seem to be panicked by humans coming near them—they may squawk hysterically, or retreat to the back of the cage. These hens, despite the crowding, were lively, active, and showed no fear of us. They started pecking at our boots and grabbing the shoelaces with their beaks. Jesse picked one up, cradling her to his sweatshirt with a hand around her breast. She sat there calmly, looking at us. "They each have a personality," he said.

We could see that the hen Jesse was holding had had the end of her beak cut off. "We tried not debeaking," Jesse said, "but it didn't work for us, we had a lot of trouble with feather-pecking." Some breeds, he explained, are more aggressive than others. That leads them to peck each other. If the pecks draw blood, other hens will gang up on the unfortunate victim, sometimes pecking her to death. With the less aggressive breeds they are using now, mortality is down to 1 percent. If they can find a breed that is sufficiently non-aggressive and still lays well, Jesse said, they might see if they can manage without debeaking.

The most crowded area of all was near some tubular structures that stretched down the sheds, one on each side. These were the nesting boxes, offering the hens some privacy in which to lay their eggs. They were eager to get into them. Around one nesting box there was a particularly large crowd of hens, shoving each other aside and jockeying for position to enter. When we asked Jesse about that, he laughed and said that hens are like fans at rock concerts in that they have a mob mentality. They will crowd all over each other to get into a particular nesting box, although the one right next to it—which is identical, as far as he can tell—is empty.

If the cost of eggs from uncaged hens is to be kept within reasonable bounds, it is crucial that the hens lay their eggs in a nesting box, because there is equipment under the nesting boxes that automatically carries the eggs to the front of the shed for collection and processing. If the hens lay their eggs on the floor, the eggs have to be collected by hand, and in so crowded a shed they wouldn't be easy to find and are much more likely to be cracked or broken. But Pete and Gerry's hens like their nests—less than 1 percent of the eggs are laid on the floor.

When we had looked as much as we wanted to, we moved out of the shed to continue our conversation away from the cackle of 20,000 hens. We asked Jesse how long the farm had been producing organic eggs.

"About eight years. It really saved the family farm. We wouldn't be here otherwise. My grandfather started this farm. He could make a living with 5,000 hens. Nowadays, in the conventional industry, a million hens is a small farm. We never got to that enormous size, never really wanted to, but we found we were just getting pushed out of markets everywhere. It's get big, or get out. Luckily, my father found this niche and it was the birds that got out—of the cages. We have about 100,000 hens now, in six sheds. That makes us one of the larger organic egg farms, and it's enough to make it work, economically."

Organic egg producers get, Jesse said, about four times the price paid for conventional eggs. Their costs are higher, of course. The shed we were looking at could have held about 150,000 hens stacked up in tiers of cages. The organic grain used as feed costs nearly three times as much as conventional grain. Apart from the grain, the only thing the hens are fed is organic flax seed, for its omega-3 content, which is thought to have health benefits.

We discuss how the organic certification works. USDA, the United States Department of Agriculture, sets the standards, but doesn't actually do the certification itself. There were certifiers already in existence when the national standards were set, so USDA trained and accredited them. Some of the certifiers are non-profit organizations, some are for profit, and in some states—New Hampshire is one of them—the certification is done by the state department of agriculture. That can make for different interpretations in different states, however. The most controversial rule, Jesse told us, is the requirement for outdoor access for the hens.

We gave each other quizzical looks. We hadn't noticed any way in which the hens we had just seen could go outdoors. "So these hens have outdoor access?" we asked.

Jesse pointed to a bare patch of dirt between the shed we had been in and a neighboring shed. "There are penned in areas over there, and around the back," he said. But the shed we had been in didn't seem to allow any way of getting outside. We asked Jesse how the birds got out. "It's sealed," he acknowledged. "I sealed it up about three or four weeks ago, because of the time of year. The USDA has exceptions for that, depending on the climate. Theoretically, an organic farm in Arizona ought to be able to give the birds access to the outside all year. But the controversy is really about disease, if the hens come into contact with wild birds."

We knew about avian influenza, or bird flu, which has led to more than a hundred million chickens dying or being slaughtered in Asia and caused some human deaths. We asked if you couldn't net the area so that wild birds would not come into contact with the hens. Jesse said that was possible, but birds flying overhead could still infect the hens through their droppings: "Even though it's a remote possibility that a duck flying over would happen to be carrying avian influenza, if it did drop feces, the hens would pick it up. One would get the disease, and then when that bird came back into the shed at night, they'd all get it. Then the government would probably come in and exterminate the entire flock. So it's a real fear. The way they wrote the USDA rules about access to the outdoors, they were for cattle, and they haven't changed anything for hens. I've got letters in my office from veterinarians and poultry experts saying, 'Do not let your hens outside.' It won't happen today, it won't happen tomorrow, but at some point, having all those hens out there, exposed to wild birds, they're going to get something."

There seemed to be a clash between the rules for producing organic eggs and what Jesse believed was necessary to protect his hens and his livelihood. We could see the problem, but we weren't clear how Jesse was resolving it. We pressed on: "If we had come a month ago, and it was a warm day, would there be hens out there?"

"If it was a clear day, and we could be sure that there were no wild birds flying over, yeah."

"But you can never be sure that there would be no wild birds flying over!"

"Right. There's the rub. But we've had the doors open on some days. Not many birds go out."

"And you don't get problems with the inspectors about that?" we asked.

"No, nobody is getting problems about that, at this point."

## CERTIFIED HUMANE

We moved on to the other form of certification Pete and Gerry have—Certified Humane. This means that the product meets a complex set of standards set by Humane Farm Animal Care, an independent nonprofit organization supported by two of the largest animal welfare organizations in the country, the Humane Society of the United States and the American Society for the Prevention of Cruelty to Animals. The standards for laying hens run to 22 pages. They require that hens have enough space to stand normally, turn around, and spread their wings, and to perch or sit quietly without disturbance. That rules out the usual cages, of course. Hens must also have nest boxes, litter to scratch around in, and they must be able to "dust bathe," getting down on the litter or dirt to fluff dust up around their feathers. This seems to be an important need for hens, perhaps because it helps them to control parasites. The hens must have access to food every day, so they cannot be forced to molt by starvation. Other rules set standards for the quality of food, water, and air; for keeping records; for inspecting the hens; and for the way in which "caretakers" are trained to handle the birds.

Despite being strict in some respects, the standards are intended to be commercially realistic. They do not require that the hens be able to go outside. They permit what the standards call "beak trimming," more accurately described as beak searing, "where there is a risk of outbreaks of cannibalism." (If alternative ways of preventing cannibalism are discovered, beak searing will be prohibited.) A more serious deficiency in the standards is that they say nothing about how the hens are transported to slaughter, nor about how they are slaughtered. Often "spent hens"—as the laying hens are called when they have reached the end of their productive lives—suffer a fate even worse than meat chickens, because they are not good to eat and have little value. Some producers have been known to pack them into containers and bulldoze them into the ground—burying them alive. In 2003 a San Diego County, California, egg producer was investigated by the county's Animal Services Department because a neighbor saw his workers emptying bucketfuls of squirming hens into a wood chipper. The farm owners acknowledged getting rid of 30,000 hens by that method. They were not prosecuted because the Animal Services Department concluded that they were "just following professional advice" from two veterinarians. The department did, however, lodge a complaint against one of the veterinarians, Dr. Gregg Cutler, who denied directly authorizing the use of the wood chipper in that case, but condoned the method of getting rid of spent hens. Dr. Cutler is no rogue backwoods vet: he is a member of the animal welfare committee of the American Veterinary Medical Association. More recently, in 2005, a Missouri Moark producer was found disposing of thousands of live hens in a dumpster.[1]

Jesse told us that he is really happy to be working with Humane Farm Animal Care, and, especially, that his hens are not in cages: "I grew up with seeing chickens in cages, and I wouldn't be in chickens if we had to do that. It's really nice to see them like this... to be able to see the chickens interact."

He asked us to tell him what we thought of the system. We told him that we had been surprised to see so many birds so densely packed into one shed. It wasn't what we'd imagined an organic farm would be like. He replied that he'd like to give the birds 2 square feet each, but a higher density of birds generates more heat, so that providing more space would mean a bigger heating bill in winter. That, plus the fact that he would need more sheds to get the same number of eggs, would throw out the economics. "It's not idyllic, but it's very, very much better than the cages, and it's the best we can do with the market situation we have."

Before we left, we couldn't resist asking who "Pete and Gerry" are. Were they copying Ben and Jerry? No, Jesse said. Gerry is his father, the one who got rid of the cages. "Pete still works on the farm too. We're the real thing."

Driving back through the glorious New Hampshire scenery, we discussed what we had seen. The birds seemed reasonably contented and looked much better off than caged hens. But we were disturbed by the fact that there were so many of them in a single shed, effectively unable to go outside, and certainly never able to enjoy scratching around in grass, or to be part of a normal-size flock in which they get to know each other as individuals. And of course, after 56 weeks of laying, these hens were going to be sent off to be killed. Hens commonly live for more than five years, and some have passed ten years of age, but after just one year of laying, hens start to lay fewer eggs, and it becomes uneconomic to keep them. By inducing them to molt, their productive life can be extended, but only by another few months.

## BETTER OPTIONS?

In the weeks after we visited Pete and Gerry's Organic Eggs, we asked around for really free-range eggs—where the hens are actually out on grass, at least during the warmer months and have not had their beaks seared off. It wasn't easy to find commercial suppliers of such eggs in the United States. We talked to Cyd Szymanski, known in the Denver area as "the Egglady" for her vigorous personal promotion of "eggs with a conscience." Szymanski comes from the family that owns Moark Eggs, one of the largest producers of battery eggs in the United States, selling more than $500 million dollars worth of eggs a year, produced by more than 24 million birds, virtually all in cages. (Moark, now 50 percent owned by Land O'Lakes, sell their eggs under various labels, including "Country Creek," the label on the carton bought by Jake Hillard at Wal-Mart.) Szymanski's business is rather different: "We've never had any caged birds and we never will." That way, she says, she can market "eggs with a conscience." Even so, all her birds—about 350,000 of them—have their beaks cut back, and only the hens that lay her organic eggs get to go outdoors. In spring, these hens even have some grass to scratch around in, until the dry Colorado summers kill it off.

Organic Valley, a co-operative of small farmers, has for several years had a policy requiring outdoor access for hens, weather permitting. Their producers have to offer pasture amounting to 5 square feet per hen, and it cannot be all bare dirt—pastures must have some grass or alfalfa cover. Nick Levendoski, who coordinates Organic Valley's pool of egg producers, says that 90 percent of them have at least half of the outdoor area covered with pasture. The policy "created quite a stir out there" when it was implemented a few years ago, Nick told us, but the producers took a "leap of faith" and did it, and the feedback has been "nothing but positive." The producers—there are 60 of them, with flocks ranging from 600 to 16,000 hens—feel that giving the birds outside access has "enhanced the overall bird health, enhanced their production." Although most of the farms are near the Mississippi River, a prime flyway for wild birds, and almost all of them have streams with waterfowl on or near their properties, there have been no major disease outbreaks. Organic Valley does, however, permit debeaking.[2]

Eventually we heard about some farms that do not debeak their hens. Grazing Fields is a co-op of about a dozen farmers around Charlotte, Michigan, who, in total, have about 4,000 hens and produce 3,000 eggs per week. They sell mostly to health food stores and restaurants in the Detroit area. We also spoke to Mark Tjelmeland, who sells "TJ Family Farms" eggs to stores in Ames, Iowa. His hens can go out to pasture when they want, and they all have intact beaks. Over ten years, he says, he has only had one flock that started to peck each other.

Commercially speaking, these are all very small producers. But in many farmers' markets across the country, one is likely to meet local farmers who produce small quantities of free-range eggs from hens with intact beaks. Ask around, as farmers don't always bring all their goods to market. And ask how the eggs are produced, or even visit the farm for yourself. In areas without such markets, there are other options. Some of the most animal-friendly eggs on sale in the United States come, believe it or not, from New Zealand. On the temperate, well-grassed plains just south of Auckland, where hens can be outside 365 days a year, Graeme Carrie runs a company known as Free Range Eggs New Zealand, or Frenzs. He buys eggs from several farmers, who have flocks of about 1,000 birds each. None of them debeak their hens. "It's an ugly process," he told us, "The beak was designed for them to eat and range properly. If you debeak them, they can't do that. It's as if you chopped off the tips of your fingers, you wouldn't be able to pick up things well then, would you?"

Although there is strong demand for Frenzs eggs in New Zealand,

Frenz sends several thousand dozen eggs every week to the United States. Initially the eggs were sold only in Los Angeles, which has the most direct and frequent flights from New Zealand, but now they are also available throughout California and in Nevada, Arizona, Colorado, and Washington. So despite the distance, they can be on sale within 24 hours of being laid—in fact, because New Zealand is 20 hours ahead of California and the flight takes only 11 hours, Carrie jokes that they can be in the store before they are laid.

In Britain, bodies like the Soil Association and the Biodynamic Agricultural Association (Demeter) have set standards higher than the European Union's standards for organic egg production. For example, where the EU regulations allow flocks of up to 9000 chickens to be housed in a single shed, the Soil Association allows a maximum of 2000, and in many cases, sets a limit of 500. The Soil Association and Demeter also have higher standards for the quality of the pasture, setting longer intervals during which the grass must be rested before chickens can be returned to it.[3]

## WHAT PRICE GOOD EGGS?

We could see the dilemma, though. Jake paid $1.08 for a dozen Country Creek eggs at Wal-Mart. Mary Ann paid $1.69 for half a dozen Pete and Gerry eggs. Frenzs eggs, whose price includes trans-Pacific airfare, retail for between $4 and $5 per half dozen. Carrie acknowledges that people can buy much cheaper eggs than his, but he's not prepared to make his eggs cheaper by depriving his hens of the freedom to enjoy fresh air and sunshine and peck at grass, seeds, and grubs. He thinks that great-tasting eggs are good value, even at 70 cents apiece. Cyd Szymanski, whose "eggs with a conscience" sell for $2.99 a dozen, says that when people complain about the cost of her eggs, she says: "Look, you can't spend an extra ten cents an egg, or twenty cents an egg, to make the chickens' lives better? You can't spend 20 cents an egg and then you go out and you buy a $4.50 latté? And it's not even food!" That makes sense—and compared with the price people pay for a latté, even 70 cents is not a lot for an egg. But the leap from 10 cents an egg to 70 cents an egg is simply too great for most people to take. As long as the battery egg system is with us, it is valuable that egg buyers should have another option that makes it easier for them to move up to eggs from hens who are not kept in cages, even if they cannot go outside.

We also thought it was a plus that Pete and Gerry, like Cyd Szymanski, Grazing Fields, and TJ Family Farms, have no hens in cages. As the sales of organic and cage-free eggs have increased, most of the big battery egg producers have sought to tap this new market by setting up a separate line of "cage-free" eggs. Cyd Szymanski questions the ethics of people who "have got millions of birds in cages, but they'll put down a unit with 100,000 cage-free birds." She tells us that at times these producers will have surplus caged eggs they can't sell, and asks us: "Where do you think those eggs are going to be going?" Then she answers her own question: "In cage-free cartons." It is hard to say how often that happens, but we do think that anyone who eats eggs and cares about the way hens are treated would do best to support producers who refuse to have any hens in cages.[4]

That still leaves a different question: If something like Pete and Gerry's mode of keeping hens is the best system of producing eggs that most people are prepared to pay for, would it be better to follow the example of the Farb family, and eat no eggs at all? We will be in a better position to consider that suggestions after we have seen what the Farbs eat.

# 9

# SEAFOOD

Seafood is a significant, and growing, part of the diet of most American families. Tuna was for many years the favorite American seafood, but it has now been surpassed by shrimp. Salmon is also a very popular choice. But many fishing stocks are in decline, and some—including Atlantic salmon—are commercially extinct. Cod, for centuries an icon of New England, has become so scarce that there is a running joke among Massachusetts fishing people that Cape Cod will have to be renamed.

Jake Hillard bought Gorton's breaded fish fillets at Wal-Mart. Mary Ann Masarech's seafood purchases were Norwegian Salmon fillets, Maryland crab cakes, and shrimp kebobs, all bought online from Horizon Foods and delivered to her door. These different purchases involve very different kinds of sea creatures coming from very different waters. They raise many different ethical issues, but we can group them into two broad kinds: environmental issues, and concern about animal suffering.

## THE TRAGEDY OF THE COMMONS AND THE STORY OF COD

Our love for fish is wiping many of them out. World consumption of fish and fishery products—the "products" include fish meal for animal feed and fertilizer—has more than tripled since 1960. A quarter of the world's commercially important ocean fish populations are depleted or slowly recovering from past over-exploitation; another 47 percent are fished to the full extent of their capacity, so that increasing the number of fish caught would risk causing a population collapse. Worldwide, we are eating around 100 million tons of "seafood"—marine life—a year. Americans eat, according to one estimate, 17 billion marine creatures a year.[1]

Commercial fishing methods have become at once more efficient and more wasteful. Bigger boats and bigger nets capture greater numbers of fish than ever before. But their gear damages the seabed and scoops up unwanted species—officially, "bycatch," but at sea, more bluntly termed "trash" and just thrown overboard, usually either dead or dying. In some fisheries, the bycatch ratio is outrageous: shrimp trawlers take many times more tons of bycatch than they do of shrimp. Each year about a quarter of all fish taken worldwide is bycatch—that's some 27 million tons, billions of living creatures, trashed. In U.S. fisheries, the bycatch is 22 percent, or 1.1 million tons, which according to Dalhousie University Professor Ransom Myers is enough to fill every bathtub in a city of 1.5 million people.[2]

The problem with commercial fishing, from an environmental perspective, is that each fishery in international waters is a commons, and in a world of self-interested independent agents, as Garrett Hardin argued in a celebrated article published in 1968, the tragic economic logic of the commons rules.[3] Imagine a village that owns some common land on which, traditionally, every family in the village has the right to graze their cows. In the olden days, every family in the village had just one cow, because one cow would provide the family with plenty of milk, butter, and cheese. The common provided ample grass for that number of cows. But as roads improved, some families realized that they could sell extra cheese to a nearby town, so they put more cows on the common. After a while, it became clear to everyone that there were too many cows for the common to support. Soon the cows were not eating well and were giving less milk. The families tried to make up for the lower yield per cow by putting even more cows on the common. Each family was following its own interests in trying to get as much out of the common as possible. Because some families could not imagine doing without the extra income they received from selling their cheese, attempts to agree on a limit to the number of cows each family could graze broke down. At that point, even families that would have been willing to accept limits said to themselves: "There are too many cows on the common. But if we take our cows off the common, we will lose income, and cows belonging to other families will eat the grass that our cows would have eaten. Why should we do that?"

There was no faulting the logic of the reasoning of each family, but the outcome of this pursuit of individual self-interest was easy to foresee. In the absence of any authority with the power to restrict the traditional rights of the villagers, the common became a barren, eroded field, the cows starved, and all the villagers were worse off than they would have been if only they could have controlled the right to graze on the common.

Change grass to fish, and we have the story of the cod. When John Cabot discovered the Grand Banks off Newfoundland in 1497, he reported that codfish ran so thick you could catch them by hanging wicker baskets over a ship's side. A century later, fishing skippers were still reporting cod schools "so thick by the shore that we hardly have been able to row a boat through them." Feeding on the rich nutrients created by the meeting of the cold Labrador Current with the warm Gulf Stream, the cod grew big, some reaching six and seven feet long, weighing as much as 200 pounds.[4] Because they tend to congregate in great, dense schools, they were easy picking for fishing boats from both sides of the Atlantic and fed nations for centuries.

In the 1950s, factory ships—floating processing plants equipped with filleting machines, fish-meal grinders, and great banks of freezers—came to the cod waters. Their radar enabled them to find and haul up whole schools of fish, and they stayed out fishing and processing around the clock for weeks on end. The ships came from all over the world. They became bigger and hauled nets with mouths over a thousand feet wide that could haul in as much cod in an hour as an early cod fishing boat would catch in an entire season. Many of the captains could see that the cod could not sustain such catches indefinitely, but the fish were in international waters, and no one, it seemed, could stop the plunder. The captains knew that if they did not catch the cod, someone else would.

The looming disaster gave rise to a desperate attempt to save at least some of the cod fishery. In 1972, Iceland unilaterally extended its territorial waters and announced that it would control fishing. That brought Britain and Iceland close to war. Icelandic gunboats threatened British trawlers that did not observe its new limits, and the Royal Navy was sent out to protect the British ships in what were, Britain maintained, international waters. But Iceland's actions made good sense, and British public opinion would not support war with the plucky Viking descendants, so an agreement was reached that accepted Icelandic authority over the fishing grounds but allowed some British fishing to continue. Canada then followed Iceland's example, extending its territorial waters in 1977 and effectively closing the cod fisheries to other nations. Alas, Canada allowed its own fishing ships to continue the slaughter right into the 1990s. By then the cod populations had already crashed. Surveys in 1992 indicated that the adult cod population was a mere 1.1 percent of that in the early 1960s. That year Canada finally closed the fisheries entirely. More than a decade later, the cod have not recovered. Perhaps they never will.

Similar stories, if not so vast in scale, can be told of fish stocks in other parts of the world. Declining catches of species after species have alerted governments and conservation groups to the problem and the need to move toward more sustainable fisheries. One outcome is the Marine Stewardship Council, formed in 1997 by the World Wildlife Fund and Unilever, the world's largest buyer of seafood. Now an independent nonprofit organization, the Council has worked with industry and conservation organizations around the world to establish an environmental standard based on the UN's Food and Agriculture Organization's Code of Conduct for responsible fisheries. Any fishery may apply to be assessed against the Council's standards. The Council then engages independent organizations to assess whether the fishery is being carried on in a responsible manner and is sustainable. The Council has given certification to fifteen fisheries to date, entitling them to label their products with its blue "Fish Forever" seal.

Although sometimes criticized by environmental groups for being too soft on the fishing industry, the Marine Stewardship Council's standards are a step toward sustainable fishing around the world. They are based on three principles:

1. **The condition of the fish populations.** Are there enough fish to ensure that the fishery is sustainable?

2. **The impact of the fishery on the marine environment.** What effect is the fishery having on the immediate marine environment, including non-target fish, marine mammals, and seabirds?

3. **The fishery management systems.** What are the rules and procedures in place and how are they implemented to maintain a sustainable fishery and to ensure that the impact on the marine environment is minimized?

## Gorton's Fish Fillets

How do Gorton's "breaded fish fillets," eaten by Jake, Lee, and their children, measure up against these standards? The fillets, we learned, are made from pollock, a cousin of the cod that swims the icy waters of the Bering Sea between Alaska and Siberia. The U.S. fishing fleet dominates the pollock fishery, landing about 1.5 million metric tons—or 3.3 billion pounds—each year. Since pollock usually don't weigh much more than 2 pounds, this means at least 1.6 billion pollock are taken every year just from this one region. If you had an entire football field to cover with fish and piled up all these fish on it, they'd make a stack over 1,100 feet high—or almost as high as the Empire State Building.[5]

Despite the vast number of fish caught, the Bering Sea and Aleutian Islands pollock fishery became, in October, 2004, the eleventh fishery in the world to be certified under the Marine Stewardship Council program, and the smaller Gulf of Alaska pollock fishery was certified the following year.[6] Pollock fishing, it seems, does relatively little damage to the marine environment of the northern Pacific. Most pollock are caught by "mid-water trawl" nets and processed at sea. Smaller numbers are caught by "longline" gear. Neither of these methods uses gear that touches the ocean floor, where it could damage sensitive environments. Because pollock swim in large, dense schools, bycatch is low. Seafood Watch, a program of the Monterey Bay Aquarium, gives pollock a "best" rating. Environmental Defense's Oceans Alive Web site says that pollock stocks are "in good shape" and well-managed, but mentions a concern about the removal of pollock that are a food source for the endangered Stellar sea lions.

When we called Gorton's to ask if their fish fillet packaging bore the Marine Stewardship Council's "Fish Forever" seal of approval, their consumer affairs representative appeared not to have heard of the program or the pollock fishery's recent approval. No seal or other statement on the packet indicates anything about whether the fish it contains is sustainably fished. If Gorton's does not advertise and market the environmental status of its own product, Jake presumably made a relatively environmentally friendly choice more by luck than by design. Like most consumers, she bought the fish fillets for taste, price, and convenience.

## Horizon Seafood

Jim and Mary Ann try to be ethical shoppers, but as we have seen, they are both busy people and must, as Mary Ann says, "balance ethics and convenience." That leads her to buy foods from Horizon: "This guy Kevin comes to the door every six to eight weeks. He's like a Good Humor man. He has a truck with big boxes of beef, free-range chicken breasts, and a lot of fish. It's all portion-controlled and flash-frozen, much of it natural, antibiotic-free, and no growth hormones. So it's great, because I don't have to do a thing. We have a big freezer. He shows up or I just call him and say, 'I need to order some more."

Like many conscientious omnivores, Mary Ann and her daughters tend to choose sea animals rather than land animals because they want to eat lower down the evolutionary ladder. As we saw, she thinks that a fish, unlike a cow or some other mammal, "isn't going to look me in the eye with any sensitivity." She showed us one of the Horizon Foods brochures that she uses to place her orders. Its sixteen glossy pages show colorful close-up photographs of grilled steaks, rare roast beef, crispy-looking breaded chicken "fingers," broiled fish, hams, pizzas, and many other items. She pointed out three of the foods that she regularly buys. We looked to see what Horizon had to say about them. The first was crab cakes. The caption by the picture says: "All natural Maryland crab cakes are handmade with only the finest ingredients." It goes on to give serving and cooking suggestions. In waters around the United Kingdom, the Hastings Fleet fisheries for Dover sole, mackerel and herring, the North Sea Herring Fishery, the South West Mackerel Handline Fishery, and the Thames Blackwater Herring Drift Net Fishery, have been certified as sustainable.

There is no further information about the Maryland crab or crab fisheries anywhere in the brochure. The other items were salmon fillets and shrimp kebobs. The brochure gave no information about these items other than that they are "flavorful and tender" and "tasty." We looked at the Horizon Foods Web site and found nothing further about the sources of these items. Perhaps Jim and Mary Ann could have been more diligent in researching crab, shrimp, and salmon, but with their schedules, that's asking a lot. Horizon wasn't making it easy for them to make ethical choices, and so we did some investigating of our own.

## *Crab*

Labels can be deceiving. "Maryland crab cakes" are a case in point, as Angelina's of Baltimore, a restaurant and crab cake retailer, warns:

> A restaurant can call any crab cake a "Maryland Crab Cake" because this technically only refers to the style in which it is prepared. In reality it may be a "Thailand Crab Cake." Just like anyone can call any cheesecake a "New York Cheesecake."[7]

So we checked with Horizon Foods' office in Plainview, New York, where a product buyer assured us that their crab cakes truly were from Maryland crabs—90 percent of the time anyway, and the remainder were from the Virginia side of Chesapeake Bay.[8]

That's why at 5 o'clock one August morning we found ourselves in a shack on Smith Island, off the eastern side of Chesapeake Bay, listening to the weather forecast with Eddie Evans, Sr. Eddie, 67, is a waterman who began his commercial fishing career working on a skipjack, the shallow-draft sailing sloop once common in the Bay. His son, Eddie Jr., and grandson, Craig, come in and sit down. They are the twelfth and thirteenth generations of Evans on Smith Island. The weather station is reporting winds of 15 to 20 knots. Eddie Jr. says, "It's going to be pretty rough out there." Eddie Sr. tells us that he won't be going out today because of a doctor's appointment.

We get on the *Janet Lynn*, named after Eddie Jr.'s youngest sister, and motor five miles out to the crabbing grounds off the eastern side of Smith Island. Eddie Jr. and Craig put on rubber boots, rubber overalls, and long, black rubber gloves. Craig takes his station at a set of controls toward the rear of the boat. Eddie stands just behind him at an aluminum table with high sides like a big tray. With the *Janet Lynn* pitching and rolling, Craig maneuvers up to a white float, slows the engine, and with a boat hook pulls up the float and the rope attached. He puts the rope into the groove of a spinning pulley next to his motor controls. In seconds, a crab pot bobs up at the side of the boat. In one motion, Craig whips the rope out of the pulley, grabs the pot, and swings it backward and up onto the aluminum table. Then he speeds the boat ahead to the next float while Eddie turns the pot upside down on the table. The pot is a squat cube about 20 inches on a side made of one-inch mesh "chicken wire." Low down on each side is a round opening about 6 inches across with a cone of wire that tapers to 2 to 3 inches inside. These cones allow crabs to easily crawl inside, but make it difficult for them to crawl back out. The pot has one of its four sides loosely attached. When Eddie turns the pot upside down and shakes it, the crabs fall through onto the table and skitter about in the strange, bright environment.

These are blue crabs: *Callinectes sapidus*, a combination of Greek and Latin words that translate as "beautiful swimmer that is savory." They live here and in other bays, estuaries, and shallow waters of the Atlantic Ocean from Nova Scotia to Argentina. The ones on the table wave their bright blue and red claws and snap them menacingly at an enemy they cannot grasp. Eddie quickly sorts the crabs, throwing small ones back in the water and pitching the others into bushel baskets, one each for females, two sizes of males, "peelers," and "soft shells." A "peeler" is a crab that is about to shed its shell, as crabs periodically do when they grow. Immediately after the shell has "peeled" the new shell underneath is soft, and the crab is a "soft shell." When in doubt as to a crab's size, Eddie holds one in a cut-out on a side of the aluminum table. The space is 5¼ inches—the size under which a crab must be returned to the Bay this time of year.

And so it goes, float after float, pot after pot, twenty-five pots to a "line" or "row." At the last pot of each line, we go forward into the shelter of the cabin and Eddie speeds back down to the first pot in the next line. When we hit a big wave, spray flies over the boat and falls in sheets on the working area aft. The Evans have about eight lines in this area and lines in several other areas. This morning we haul about 300 pots and take an average of three or four crabs from each one. The bycatch today is 4 to 5 small fish, each one thrown back immediately by Eddie, and a dozen jellyfish or parts thereof.

Fishing boats are coming and going all around us as we head back to the harbor. At its entrance, a large hand-painted sign on a building owned by the local watermen's association, says: "Thank you for visiting Smith Island. Please help preserve our heritage. Do not support the Chesapeake Bay Foundation." Eddie and Craig wait their turn at the buyer's dock, and then hand up three bushel baskets to be weighed and sold. They retain a few other baskets of crabs to keep in their walk-in refrigerator. These will be filled, weighed, and sold another day.

Eddie Sr. is working in his crab shack just a few feet above the water of the harbor and literally in his backyard. He is sorting out peelers and keeping them in the water circulating through a series of fiberglass tanks until they molt into soft shells. "That's where we get our most money, that soft shell." He explains that most of the biggest male crabs, six inches or so, go into the "basket trade"—crabs sold whole for stores and restaurants. Smaller crabs and the female crabs go to the "picking trade" and become, among other things, the famous Maryland crab cakes bought by Mary Ann. He talks about the economics of crabbing. There are wages for the mate, the person who goes out with you, and fuel costs, plus maintaining the boat. Then there are the buyer's costs, processing, refrigeration, and transport: "When you add it all together, then what you're paying for a pound of crab meat doesn't sound quite as bad."

We bring up the controversy about the state of the Chesapeake Bay and its marine life. The blue crab has historically supported one of the hemisphere's largest and most profitable fisheries because it grows fast, can thrive in a wide variety of habitats, is not a picky eater, and is a prolific breeder. In unexploited waters, a female mates once and carries enough sperm in reserve to produce as many as 14 million fertilized eggs a year for as long as she lives.[9] Eddie says that the crab population is "on the increase" because the state is regulating traps, seasons, the size of crabs taken, and the number of licenses. "It's control heavy," he says, "and yet, it boils down to the main culprit on the whole Chesapeake Bay is pollution. If this Bay was as clean as it used to be, things would produce on a normal cycle and they wouldn't need any regulations. But that ain't the way it is, and it's not the way it's gonna be." The pollution, he says, is coming from farms, fertilizers used on golf courses, and development along the shoreline. "I wouldn't put my finger on any one thing and say this is the culprit, it's just a combination of a lot of things."

We tell Eddie about the sign we saw coming in the harbor and ask him about the Chesapeake Bay Foundation. "They've got good goals," he says. "They started out to make the public aware of pollution in the Bay and all, but now they've more or less come to be a regulatory thing. And anytime you get environmental groups, they don't like anything you do. They want to conserve everything." Eddie's calm demeanor is giving way now. "If they preserve that, we'd starve to death as far as that goes. I always said that a waterman is actually an environmentalist. We don't mistreat our environment, we don't abuse things, because that's our living. But yet, there's a line you can cross there and be too environmentalist. And I don't like to lump everybody in the same basket, because some's got good commonsense, but you got a certain amount of them out there that would conserve just to conserve. They think Sam Walton produced it all, out of them factories and stuff. I got news for them, it either grows on the land or it comes from the water."

We checked with the Chesapeake Bay Foundation for its views. The Foundation dates from 1967 and is the largest organization dedicated solely to conservation of the Bay's waters and a watershed that sprawls across six states and the District of Columbia. Bill Goldsborough, Senior Fisheries Scientist, said of the protest by Smith Island crabbers, "We are not radical environmentalists by a long shot." He explains that the vast majority of their work is on pollution reduction, but it is work the crab fishermen don't see. "They only see the things we do and say concerning fisheries, and they react to that." That work puts the Foundation at odds with Evans' belief that the crab population is increasing. "That doesn't agree with the science on it," he says. "That's probably his observation locally, and that's very valuable, but you have to put it in context and realize that it's just one data point." The states of Maryland and Virginia conduct several surveys each year. Particularly useful, according to Goldsborough, is the winter dredge survey in which samples are taken at over a thousand locations around the Bay when the crabs are dormant and the population is easier to measure. Those surveys indicate that blue crab populations in Chesapeake Bay are in decline.

The Foundation is in conflict with some watermen in other parts of the bay because of a dramatic increase in the number of peeler and soft shell crabs taken. As Evans told us, these immature crabs are particularly profitable, so crabbers have an incentive to take them before they have matured and mated, and this puts more pressure on the population to maintain itself. (The Smith Islanders, Goldsborough told us, have not significantly increased their take of peelers and soft shell crabs.) But even if the decline in the crab population is solely due to pollution problem and not overfishing, the Chesapeake Bay Foundation believes that responsible stewardship requires everyone to be more conservative with the taking of crabs, and that makes crabbers feel victimized. "We are a lightning rod," Goldsborough says. "It comes with the territory."

Eddie is concerned, too, about the fishing industry's diminishing opportunities for young people like Craig, his grandson. The problem, as he sees it, is that it is becoming too expensive to get started and the earnings are falling. He rattles off some figures and concludes that Craig would probably be "way ahead of the game" working for a company rather than becoming a crabber. Smith Islanders, whose simple lifestyle centers around church, family, and the water, are concerned that people are moving to the mainland. From a high of about 700 in better times, only about 300 now live in the island's three villages. Among other forces are the high prices paid by mainlanders who want second homes in a quaint, seaside setting. Home prices, some told us, had "quadrupled" in the past few years. "When the watermen keep getting off the water and they turn it into a playground out here, then what's going to happen? Where's people going to get their seafood?"

With the demand for crabmeat outstripping the domestic supply, U.S. processors now import a related species, *Portunus pelagicus*, commonly named "blue swimming crab" and "flower crab." About three-fourths of the crabmeat products sold in the U.S. contain crabmeat imported from Indonesia, the Philippines, Thailand, India, and other countries where the labor of crabmeat pickers is cheap and there may be little regulation to ensure sustainable fishing practices. We ask Eddie what he thinks about this: "Yeah," he says, "people in America, they might see a can of crab meat on the shelf up there four or five dollars cheaper than they can buy it here in America, but do they know what they're eating out of that can? How it is produced and how it's packaged? They don't know."

Eddie is right about that. In the Philippines, a World Wildlife Fund study found that *Portunus pelagicus* was threatened by over-harvesting, habitat destruction, and inadequate marine policy and enforcement. Motorized boats using gill nets have become the dominant method of catching crab, replacing the traditional crab pots. The study estimated that the combined length of all the gill nets in the study area alone, if tied together, was 680 miles—enough to reach from Portland, Maine, to Richmond, Virginia. Gill nets are also sometimes lost, and since they are not biodegradeable, they continue to kill large numbers of crabs who become entangled in the nets. In just ten years, overfishing has led to a 70 percent decline in the commercial catch.[10] There has been a decline in the amount of crab produced in Indonesia and Thailand too, probably also from uncontrolled overfishing.[11]

Since crab imported from Asia is often sold as "blue crab," it is not easy for American consumers to know if they are buying crab that has been caught in a sustainable manner. One of the biggest companies, Phillips Foods of Baltimore, markets "Maryland style" crab cakes made from these imported crabs.[12] Organizations such as the Blue Crab Marketing Alliance are encouraging the labeling of domestic crab with a "U.S.A. American Blue Crab" seal to differentiate them from Asian blue crab. Generally speaking, crab from the U.S., Canada, and Australia comes from managed fisheries that are likely to be sustainable. Crab imported from other countries is probably not and is better avoided.

## *Salmon*

If the Norwegian salmon "filets" Mary Ann bought from Horizon came from Norway, they were almost certainly from fish farms. Wild Atlantic salmon has largely gone the way of the cod. So few survive that 300 farmed fish are sold for every one that was caught swimming freely.[13] Most shoppers don't know this—and it is hard to blame them for their ignorance. Tests carried out for *The New York Times* in March 2005 showed that six out of eight New York City stores, including gourmet haven Dean & DeLuca, were selling farmed salmon labeled—and priced—as the far more expensive wild salmon. The difference can be detected in the laboratory by the presence of an artificial coloring fed to farmed fish in order to turn their otherwise grayish flesh pink. The flesh of wild salmon is naturally pink because of the krill they eat.[14]

Fish farming is the latest agricultural revolution and the fastest growing form of food production in the world. In 1970 it contributed only 3 percent of the world's seafood. Now about a third of the fish and other seafood we eat is farmed; the weight of farmed fish produced exceeds that of the global production of beef.[15] Almost all of this is highly intensive production. In the fjords and coastal inlets along the coast of Norway, Britain, Iceland, Chile, China, Japan, Canada, the United States, and many other countries, cages or nets that may be more than 200 feet long and 40 feet deep have been lowered into the sea and secured to platforms from which workers feed the fish. With salmon, 50,000 fish may be confined in each sea cage, at a stocking density that is equivalent to putting each 30-inch salmon in a bathtub of water.[16]

As with chicken, intensification has brought down the price of what was once a luxury food. Freshwater fish are also reared in fish farms, on land. The Bush administration has proposed new legislation that would allow offshore aquaculture and could, it says, quintuple the yield of aquaculture by 2025, by which time, it is forecast, half the fish eaten in the world will be farmed. But William Hogarth, the National Oceanic and Atmospheric Administration's assistant administrator for fisheries, acknowledges that there will be opposition from environmentalists: "Aquaculture is extremely controversial, there's no question about it."[17]

Fish farming sounds like a good way of meeting the growing demand for seafood while taking pressure off wild fisheries. But that can be like thinking that if we ate more beef, we wouldn't need to grow so much corn. Professor Daniel Pauly, of the Fisheries Center at the University of British Columbia, has described the farming of salmon and sea bass as "feedlot operations in which carnivorous fish . . . are fattened on a diet rich in fish meal and oil." Writing in *Nature,* he and his colleagues point out that although the idea makes commercial sense because of the high market price for the farmed fish as compared with the price for fish meal, these operations use up much more fish flesh than they produce, and so, far from replacing wild fisheries, actually put more pressure on wild fish populations and are "largely unsustainable." The fish farming industry's demand for fishmeal and fish oil provides incentives for fleets to take millions of tons of small fish that otherwise might provide food for cod or haddock and directly or indirectly may provide much-needed protein for coastal people in developing countries. Three or four tons of this cheap fish will be made into pellets and fed to farmed salmon in order to produce one ton of expensive salmon to sell to people in the rich nations.[18]

The fishing fleets, of course, need oil to run their engines to gather the fish to feed the salmon. One irony of this method of raising fish is that, as William Rees of the University of British Columbia has pointed out, "the salmon farming industry expends large quantities of costly and increasingly scarce fossil fuel to do several jobs that wild salmon do for free, particularly foraging at sea to catch their food." Peter Tyedmers of Dalhousie University in Nova Scotia has calculated that for every kilogram of Canadian farmed salmon produced, 2.5 to 5 liters of diesel fuel or its equivalent is consumed.[19]

Another problem with farming salmon is that the waters around the sea cages and the seabed below become polluted from the concentration of fish feces and food waste that are discharged, untreated, into the sea. The World Wildlife Fund has calculated that Scottish salmon farms discharge the same amount of waste as 9 million people—almost double the human population of Scotland. Rebecca Goldberg, a biologist with the advocacy group Environmental Defense, and Rosamond Naylor, a Stanford economist, have estimated that the Bush administration's proposal for a $5 billion fish-farming industry in U.S. waters would produce as much nitrogen discharge as untreated sewage from more than 17 million people.[20] Antibiotics and pesticides are also given to the fish to reduce their incidence of disease and of parasites, and these too float freely through the nets into the sea. Although salmon farming is a major export earner for Norway, the Norwegian State Pollution Control Agency has described salmon farms as "major polluters." Greater efforts are now being made to monitor the amount fed, and the problem of waste feed has been reduced but not eliminated.

A third issue with farmed salmon is that they frequently escape when predators or storms cause holes in the nets that form the walls of their cages. In Norway the Directorate of Nature Management estimates that half a million farmed salmon escape each year—far more than the number of wild salmon left in Norwegian waters. Better farming techniques are reducing the number of escapees, but up to 90 percent of the salmon in some Norwegian rivers are now farmed fish.[21] The escapees may interbreed with wild salmon, changing the genetics of the native species. They also infect wild fish with diseases and parasites, to which the farmed fish are especially prone because they are so intensively stocked. In Norway alone, ten million farmed salmon succumbed to disease in 2001.[22] The number of wild fish killed by parasites passed on to them from farmed fish is unknown, but recent research has shown that after passing near salmon farms, young wild salmon have levels of sea lice infestation 73 times higher than they had previously. This suggests that salmon farms are a serious threat to wild fish populations.[23]

For all these reasons, Environmental Defense gives Atlantic salmon its "Eco-Worst" label.[24] Seafood Watch gives farmed salmon a red flag, noting that while some fish farms have improved their practices, "there is currently no way to tell which salmon are coming from more-sustainable farms." Seafood Watch suggests buying wild-caught salmon from Alaska, which has the seal of approval of the Marine Stewardship Council.[25] The problem is, the *New York Times* investigation suggests, that you can't trust the "wild" label either. Unless you really have a trustworthy supplier, the only safe course is to avoid salmon. But don't switch to Chilean sea bass or orange roughy, either. "Chilean sea bass" is a commercial name for Patagonian or Antarctic toothfish, which is not a bass at all. The toothfish is a large, slow-growing fish that can take ten years to reach sexual maturity and may live for 45 years. It seems that some marketing genius decided that "toothfish" didn't sound particularly tasty, and hence would not sell well, so they renamed it. Unfortunately it then sold so well that after just ten years on the U.S. market, some local stocks have been wiped out, and the entire species is at risk of commercial extinction. The impact that this will have on the unique ecology of the Antarctic waters in which the fish live is unknown. Attempts to regulate fishing have largely failed because of the difficulty of stopping the lucrative illegal fishing in the remote seas in which the fish live. The orange roughy story is similar. The discovery, in the 1970s, of commercial opportunities for catching this deep-sea fish led to a boom that has been likened to a goldrush. Stocks that had taken thousands of years to develop were fished out in two or three years, because the orange roughy is even more slowly maturing than the Patagonian toothfish, reaching sexual maturity only around the age of 25, and living for an estimated 150 years. Scientists believe that some of these local stocks—which were taken from as far as 1,800 meters below seal level—will never recover. Others, in New Zealand and Australia, are now being controlled, but it is still unclear whether they will prove sustainable. Bycatch can equal the weight of the orange roughy caught, and fish taken from such deep waters—often from species of which little is known—are usually dead or dying by the time they are hauled up to the surface.

## *Shrimp*

As with salmon, the rapid growth of farming shrimp, or prawns, as they are known in some other countries, has led to a substantial price decline, and shrimp has taken over from tuna as the seafood most eaten by Americans. Most of the U.S. supply, whether from farms or wild-catch fisheries, is imported. No figures are available on how much of the imported shrimp is farmed, but one estimate is that it is roughly half.[26]

Domestic shrimp amounts to only 13 percent of U.S. consumption, and 90 percent of that comes from the Gulf of Mexico, generally taken by bottom-trawling boats pulling huge weighted nets that skim across the seafloor.[27] This technique can be an environmental disaster, but because the soft sediments of the Gulf's relatively shallow and storm-prone waters are regularly disturbed by natural causes, it is said to be less damaging there than in other marine ecosystems. Warm-water shrimp are short-lived and very prolific and not presently threatened by overfishing. The main environmental concern is bycatch. Even though Gulf shrimpers have equipped their nets with turtle-excluder devices and bycatch reduction devices, for every pound of shrimp caught, the trawl nets still haul up and then throw away over 5 pounds of sea creatures, including sharks and, sometimes, endangered sea turtles.[28]

The major shrimp producing countries are China, Indonesia, India, Thailand, Vietnam, Brazil, and Ecuador. These countries have little regulation of either shrimp farming or trawling, or if they do have regulations on paper, they may not be enforced. If the imported shrimp are wild-caught, unregulated shrimp trawling, without devices to prevent bycatch, will take a heavy toll on marine life. Shrimp fishing amounts to only 2 percent of the global wild seafood catch, but is responsible for 30 percent of all the bycatch in the world's fisheries. In some tropical shrimp fisheries, the bycatch is fifteen times the quantity of the shrimp caught. Thailand, the largest source of imported U.S. shrimp, is one of the worst offenders, with a bycatch ratio of 14:1.

The bycatch can include endangered species. Even if a country does legally require devices to exclude turtles, "cheating is universal" in Central American shrimp fisheries, according to Todd Steiner of the Sea Turtle Restoration Project, and it is believed to be widespread in Asian fisheries as well.[29] As a result, hundreds of thousands of turtles are killed by shrimp fishing, including endangered green sea turtles. In some fisheries, part of this bycatch may be used—sometimes even for producing fishmeal for feeding farmed shrimp. But in other fisheries, it is thrown back. Don't imagine that does the unfortunate creatures much good. As a U.S. National Marine Fisheries Service report to Congress notes, "Having been trawled up, dumped on deck, stepped on, sorted through, and left in the sun while the shrimp are collected, by the time these animals are thrown overboard, most are dead or dying."[30] Australia has substantial prawn fisheries, both in the tropical north and in South Australia's Spencer Gulf. After coming dangerously near to fishing itself to extinction, the Spencer Gulf prawn fishery has now developed a sustainable fishing practice with very little bycatch.[31] In the north, however, the ratio of bycatch to prawns has been estimated to run as high as 15:1.[32]

Apart from bycatch, the other big environmental issue raised by shrimp fishing is bottom trawling. When this is used in rocky or coral areas, the bottom-trawling nets, weighted with heavy pieces of iron to keep them on the bottom, cause extensive and irreparable damage to coral reefs and the rest of the sea floor environment, often destroying coral formations that have been built up over hundreds or thousands of years to form the environments that some species of fish and other marine creatures need to breed. Bottom trawling also stirs up sediment that makes the area unlivable for some inhabitants of the sea floor. Since it is generally done repeatedly in the same areas, it turns the bottom of the sea into something resembling a plowed field.

Shrimp farming raises a different set of environmental concerns. According to the WorldWatch Institute, nearly a quarter of the world's mangrove forests have been destroyed in the past decade, in major part to make way for shrimp farms.[33] That figure may be too high: a Seafood Watch report says that, worldwide, 10 percent of the loss of mangrove forests can be attributed to shrimp farming, but acknowledges that the figure rises to 20 percent in some areas.[34] Even sites supposedly protected by the Ramsar Convention, an international agreement signed by 141 countries to protect wetlands that are vital for the survival of migratory birds and many other species, are not safe. In Honduras, for example, La Barberie, a Ramsar-designated wetland, is being destroyed by a shrimp farming company, and the government is doing nothing about it—perhaps because the president of Honduras has business interests in shrimp farming. In Ecuador, although protests managed to close illegal shrimp farms in the Cayapas-Mataje wetland reserve, home to the world's tallest mangrove trees, shrimp farms adjacent to it are polluting the reserve's ecosystem and altering the water flow in it.[35]

The rapid expansion of shrimp farming along tropical coastlines has caused great hardship to local inhabitants. In the early 1990s, India was one of the frontrunners in expanding shrimp farming. Shrimp farmers were making good money, but hundreds of thousands of people were suffering from the pollution of coastal waters. Indian activists helped them to bring a class action against shrimp farming. Evidence presented to the court showed that for every rupee the economy gained by shrimp farming, local communities lost at least two, and in some regions, four, through damage to fishing and other resources. In 1996, the Supreme Court of India ordered the demolition of thousands of shrimp farms and awarded compensation to those whose livelihoods had been damaged by the construction of the farms.[36] Few tropical countries have a judiciary as independent as India's, however, so local villagers affected by shrimp farming elsewhere are not likely to receive even the belated compensation that Indian villagers were able to win. In Bangladesh, there have been mass protests and occasional violent clashes over the displacement of thousands of local villagers by illegal shrimp farms.[37]

Buying imported farmed shrimp may therefore be supporting a system that often does people living in developing countries more harm than good. Some shrimp farms have reduced pollution, but, as with salmon, it is, at present, impossible for the consumer to distinguish the good ones from the bad ones, and in any case, all shrimp farms suffer from the universal tendency of factory farms to feed more food to the animals than they produce for humans.

In response to our queries, Horizon said that the shrimp used in the "kabobs" bought by Mary Ann came from a shrimp farm in Venezuela, a country with a relatively small but rapidly growing shrimp industry. Environmental regulations in Venezuela are said to be among the strictest in the region.[38] Nevertheless, Seafood Watch recommends that consumers avoid imported shrimp—whether farmed or wild-caught—unless it is from Canada.[39] Oceans Alive, the Environmental Defense Web site, gives an "Eco-worst" label to all shrimp imported from Latin America or Asia—and that is, as we have seen, the overwhelming majority of the shrimp sold in the U.S., including Horizon's Venezuelan shrimp. The "Eco-best" label applies only to some less common varieties of U.S. and Newfoundland shrimp. It is possible that this generic labeling is unfair to particular countries or individual producers who have better standards than others. The problem is that it is too difficult for the consumer to make these finer distinctions and find out which producers are doing the right thing environmentally. This situation applies in the U.K. too, because almost all prawns sold in the U.K. are imported—in 2000, imports amounted to nearly 78,000 tons, whereas the domestic catch was only about 2000 tons.[40]

## The Harm of the Harvest

In addition to the environmental concerns involved, farmed fish, like Mary Ann's salmon, could suffer welfare problems similar to those of intensively farmed chicken. The fish are very densely stocked, which forces them to crowd much more closely than they would in a natural school. In a handbook for fish farmers, S. D. Sedgwick writes:

> Salmon are animals genetically programmed to spend most of their lives swimming freely through the oceans. We now confine them in tanks or cages in close proximity and frequent physical contact with thousands of others. In the open seas they would probably never have come as close to any other fish of their own kind before returning to spawn.[41]

Thirty years of breeding salmon for farming have done little to change their instincts. In a manner reminiscent of the endless pacing of tigers in small bare zoo cages, farmed salmon typically swim as a school in circles around their cage. This seems to be a response to their inability to act on their instincts.

Large fish will bully and sometimes eat small fish. To stop this, as farmed fish grow they are sorted for size, so that the faster-growing ones are separated from the slower-growing fish. The sorting takes place between three and five times during the rearing process and involves netting or pumping the fish out of their cage so that they drop through a series of bars that only allow progressively smaller fish through. The sorting adds to the stress endured by the salmon. In general, their crowded confinement gives rise to stress, abnormal behavior, sea lice infestations, abrasions, and a high death rate. Those that survive the rearing process are normally starved for 7–10 days before slaughter, to empty the gut and reduce the risk of contamination of the flesh when the fish is gutted. Any conscious being suddenly used to receiving plenty of food at frequent intervals will suffer if the food is suddenly cut off.

Then comes slaughter. There is generally no requirement for stunning or humane slaughter of fish, so they are killed in brutal ways that would be illegal and shocking if used on cows or pigs. Farmed fish may simply be allowed to suffocate in the air. It can take 15 minutes for them to die by this method. Large fish like salmon may be bashed on the head with a wooden bat, which does not always kill them outright and may just leave them injured, to be cut open while fully conscious. Or they may be stunned by the use of carbon dioxide in the water. This causes them to thrash about for half a minute, after which they stop moving but do not lose consciousness for several more minutes. They then have their gills cut and bleed to death. It is possible that they are conscious during this process.[42]

*(continued on page 132)*

## DO FISH FEEL IT?

In the popular CBS television series *Judging Amy*, there is an episode in which Amy's daughter, Lauren, tells her mother that she has become a vegetarian. Later we see the family eating dinner. Amy has cooked, and she tells Lauren, "It's ravioli. There's no meat." It turns out, however, that the dish contains shrimp. Lauren is furious, saying that her mother has been trying to trick her. Amy replies, "Shrimp is not meat," and adds, "A lot of vegetarians eat seafood."[43] For many vegetarians, that episode will have struck a familiar chord. It happens so often: you walk into a restaurant and ask what vegetarian options they have and they start telling you about their fish. It's true that some people who call themselves vegetarians do eat seafood, although that is neither the original nor the most usual meaning of the word. Mary Ann's comment that a fish lacks the sensitivity of a cow or other mammal probably sums up the reasons of many for drawing a line between meat and fish, shrimps or oysters.

Indifference to the suffering of fish is widespread even in societies where most people show some concern for animals. Otherwise, how could people who would be horrified at the idea of slowly suffocating a dog enjoy spending a Sunday afternoon sitting on a riverbank dangling a barbed hook into the water, hoping that a fish will bite and get the barb caught in its mouth—whereupon they will haul the fish out of the water, remove the hook, and allow it to flap around in a box beside them, slowly suffocating to death? Is it because the fish is cold and slimy rather than warm and furry? Or that it cannot bark or scream? Or is it because Mary Ann and people who fish are justified in believing that fish are not nearly as sensitive as mammals?

*Reviews in Fisheries Science* is an obscure journal that rarely gets noticed by anyone outside a narrow circle of scientists specializing in fisheries. But when, in 2002, James Rose, a professor of zoology and physiology at the University of Wyoming in Laramie, published in its pages a review of what we know about the brains of fish and concluded that fish cannot feel pain, his findings were reported around the world. Awareness of pain, he claimed, requires activity in very specific regions of the cerebral hemispheres—regions that fish, because of their different evolutionary history, do not possess. Fish do have nervous systems that enable them to respond to noxious, tissue-damaging stimuli that would be painful in us, but these reactions, Rose said, do not imply consciousness.[44] Anglers, not surprisingly, welcomed that conclusion.

Rose's view did not go uncontested. The following year *Proceedings of the Royal Society*, the journal of one of the oldest and most respected scientific bodies in the world (Sir Isaac Newton once served as its president) published an article by Dr. Lynne Sneddon and other scientists at the Roslin Institute and the University of Edinburgh.

Sneddon and her colleagues injected bee venom and acetic acid into the lips of captive rainbow trout and found that they rubbed their lips into the gravel at the bottom of their tank and performed a rocking motion that is common in mammals who appear to be in pain. Other fish, in a control group, had only saltwater injected into their lips and did not show the same behavior. In general, the researchers said, the trout showed "profound behavioral and physiological changes . . . comparable to those observed in higher mammals." These changes went far beyond simple reflex responses. Moreover, when the fish were given morphine, they resumed feeding, as one might expect them to do if they had been in pain and the drug relieved the pain. The researchers concluded that "fish can perceive pain." [45]

That view was supported by several other scientists, including Dr. Culum Brown of the University of Edinburgh. Brown considers it a mistake to think of evolution as a straightforward progression from simple, primitive animals to more complex, cognitively superior ones. Fish have existed for vastly longer than there have been human beings, but they did not cease to evolve when our ancestors moved onto dry land. As a result, Brown says, "The structure of the fish brain is varied and rather different from ours, yet it functions in a very similar way."

Remember the myth, given new currency in the popular movie *Finding Nemo,* that a goldfish has only a three-second memory? Brown's own research has proved that the Australian freshwater rainbowfish, at least, does considerably better. He trained them to find a hole in a net. They needed about five attempts to learn where the hole was and locate it reliably. Then he took the net away for 11 months—the equivalent, in terms of their usual lifespan, to at least 20 years for a human being. When the net was returned, the fish did not need to re-learn where the hole was—they were able to find it as rapidly as they had before it had been removed. Brown points to many other impressive cognitive feats that fish can perform, including learning from observing other fish, cooperating to catch food, and knowing their relative social rankings in a group of fish, much as chickens learn a "pecking order." In Brown's view, "we must reverse hundreds of years of prejudice" before people will appreciate that fish are intelligent. "Fish may seem quite pathetic when they are flopping around on the deck of a boat," he writes, "but get down into their world and you'll soon realise that they are remarkable creatures." [46]

We agree. As Sneddon and her colleagues have shown, fish behave as if they are in pain, and this behavior seems purposive, directed at relieving the pain—it is not merely a reflex response. We consider that Sneddon's work has, for all practical purposes, shown that fish do feel pain.

Wild-caught fish have one great advantage over farmed fish, and, for that matter, over all farmed animals. Until their final hours they live their entire lives free of human interference or confinement. This may be the best reason for thinking it to be more ethically defensible to eat fish than to eat meat. The other side of the coin, though, is that there is no such thing as humane slaughter for wild-caught fish. Each year, hundreds of millions of fish are hooked on longlines—as much as 75 miles of line, with baited hooks at frequent intervals—that are dragged behind commercial fishing boats or left in the water overnight. Once hooked, swordfish and yellowfin tuna weighing hundreds of pounds will struggle for hours trying in vain to escape. Then they are hauled in, and as they come up to the boat, fishers sink pickaxes into their sides to pull them aboard. They are clubbed to death, or have their gills cut and bleed to death.

Gill nets are another common form of commercial fishing. These nets, which can be up to a mile long, are left drifting in the sea, the top attached to floats, the bottom weighted down. The nets take advantage of the streamlined body shape of the fish, which swim into them and then are caught by the gills or fins, unable to back out. Some struggle so violently they injure themselves and bleed to death. Others remain trapped, perhaps for days, until the boat returns and hauls in the net. Then they will have their gills cut and will bleed to death, or may be left to flap helplessly on the deck as they suffocate. In bottom trawling, as we have seen, a net is dragged along the bottom of the sea, gathering up everything in its path. Fish caught in the net will be dragged along for hours, squeezed against the wall of the net by everything else that it gathers up, including rocks, pieces of coral, and other fish. Their scales may be ground off by this process. If they are still alive when the net is hauled up, those that live in deep waters may die from decompression, their swimbladders ruptured, their stomachs forced out of their mouths, and their eyes bulging from their sockets. The remainder will suffocate in the air. On factory ships that begin processing immediately, they may be cut up while they are still alive.

When we turn to shrimps, crabs, and lobster, there is more room for doubt about the capacity to feel pain than there is with fish, given the absence of a prominent brain. What is the right ethical response to such a state of uncertainty? Suppose you are driving your car along a narrow two-lane road on a dark night and you see something in the road, across your lane. It may be a bundle of old clothes, but it might also be a person. Should you just drive over it, or swerve around it, or stop? Obviously, if you can safely stop, that is what you should do. But suppose the road is icy, there is a car close behind you and a car coming the other way. In such circumstances you can't swerve, and if you hit the brakes hard you may be hit by the car behind, causing a serious accident. Then, if the object on the road is very likely to be just a bundle of old clothes, it may be right to drive over it, hoping that it isn't a person, or at least not a living one.

As this example shows, if there is uncertainty about whether what we do will cause serious harm, we should give the benefit of the doubt to the being whom we might harm. But exactly what "giving the benefit of the doubt" requires will depend on both the extent of the doubt and the costs of acting so as to make certain we don't cause harm. Similarly, if we are uncertain whether lobster, crabs, and shrimp feel pain, we should give them the benefit of the doubt and treat them as if they are capable of suffering, as long as the costs of doing so are not too high. If we are uncertain that they can feel pain, we should try to avoid doing anything that risks inflicting pain on them. On the other hand, if it were a choice between causing them possible pain or incurring great suffering ourselves, we would be justified in no longer giving them the benefit of the doubt—though we should still do our best to minimize the pain we might be inflicting on them.

If these invertebrates can feel anything, the way they are farmed, caught, and killed must inflict severe pain on them. As anyone who has walked past seafood markets knows, lobsters and crabs are often kept alive in buckets or even just piled up in a basket, with their claws tied together, for long periods. Then they are killed by being boiled alive. If shrimp can feel pain, since it takes many of them to make a meal, the suffering in every plate of shrimp could be proportionally greater than for a larger animal. For anyone who has other food choices, it cannot be ethically justifiable to risk supporting the infliction of such agony on beings who may be able to feel pain.

All of the issues we have just discussed regarding fish and crustacea can be raised about mollusks, which range from animals like octopus and squid, who have complex brains and an amazing ability to learn new tasks, like opening jars with their tentacles, to immobile bivalves like oysters. It is hard to explain what an octopus can do without assuming consciousness, but with the bivalves, the evidence for consciousness is barely stronger than it is in plants, which is to say it is vanishingly slight. Ethical arguments against eating animals that are based on not causing—or not risking causing—suffering therefore get little grip on eating oysters, clams, and scallops, but are applicable when it comes to eating octopus and squid.

## TO EAT, OR NOT TO EAT?

Despite the widespread view that eating seafood is somehow less ethically problematic than eating birds or mammals, finding seafood that is environmentally sustainable is not easy. Jake happened upon fish from a sustainable fishery, but Mary Ann, who pays more attention to environmental issues in her food choices than most people, chose Norwegian salmon, an intensively farmed fish that pollutes the fjords in which it is farmed, spreads parasites to wild salmon, and, because it must eat fishmeal, requires fishing fleets to catch three times a salmon's own weight in fish. The Venezuelan shrimp Mary Ann bought may or may not have been farmed in a way that avoids damaging its local environment, but it will still have had to be fed from fish taken from the oceans, so it is a dubious environmental choice, at best.

Of Mary Ann's three choices, only the Maryland (or Virginia) blue crab seems to have been an environmentally good choice. With the aid of an up-to-date list from Oceans Alive or Seafood Watch, it may be possible to choose sustainable seafood—although there is still the problem of fraudulent labeling of fish as "wild" when it is in fact farmed. If we add to the environmental issues the ethical obligation to avoid causing unnecessary suffering to beings who are, or may be, capable of feeling pain, it begins to seem better, as well as simpler, not to buy seafood at all, with the exception of sustainably obtained simple mollusks like clams, oysters, and mussels.

# 10

# EATING LOCALLY

When Jake Hillard bought her iceberg lettuce in the Mabelvale, Arkansas, Wal-Mart, she had no idea where it was grown or how far it had traveled. The average distance traveled by produce sold in Chicago, according to a 1998 study, is 1,518 miles. Much of it comes from California, on one of the half a million or so truckloads of fresh fruit and vegetables that leave that state every year, traveling up to 3,000 miles to reach their destinations.[1] Perhaps Jake's lettuce was on one of those trucks too. But even if the lettuce had been grown right outside Mabelvale, it would have been trucked 220 miles to Wal-Mart's highly automated 1.2 million-square-foot regional distribution center in Bentonville, moved along some of the nearly 20 miles of conveyor belts there, together with the other 450,000 cases of merchandise that pass through the center every day, and then trucked right back to Jake's local Wal-Mart. That's the nature of the distribution system used by Wal-Mart and most other national chains across the United States. The systems are designed to ensure reliability of supply rather than to minimize the distance food travels.[2]

The average distance traveled by food that is consumed in developed nations has increased, partly because international trade in food has quadrupled since 1961.[3] That increase has allowed people in the wealthier nations to enjoy foods all year-round that once had a limited season. For example, in the 1960s, North Americans ate grapes only when North American growers, mostly in California, could supply them, roughly from June through December. Now almost half of the grapes eaten in the United States are imported, many from Chile and other southern hemisphere countries, so grapes are available in the northern winter. The big increase in imported grapes naturally means that the average distance that table grapes travel to reach the U.S. consumer has also increased.[4]

The pattern is worldwide. In northern Europe, strawberries are available in January—from Costa Rica. Asparagus is flown in, off-season, from South Africa. Even chicken is now imported from Thailand, where labor is cheap. A British study calculated that the ingredients for a single meal, consisting of chicken from Thailand, runner beans from Zambia, carrots from Spain, snowpeas from Zimbabwe, and potatoes from Italy, could have traveled a total of 24,364 miles. A similar meal could have been made with ingredients traveling only 376 miles, if domestically produced ingredients were used and seasonal vegetables like cabbage and parsnips had been substituted for the out-of-season runner beans and snowpeas.[5] But domestically produced food is also traveling further, both in Europe and the United States. In the U.S., domestic transportation of grain products increased by 137 percent between 1978 and 2000, while domestic consumption increased only 42 percent.[6] Agricultural products now account for close to one-third of all domestic freight transportation.[7]

## A GROWING MOVEMENT

Mary Ann Masarech often buys her produce from Sherwood Farms, at least in the summer and fall. The farm has been owned by Sherwoods since 1713, and Tom Sherwood, who runs it now, is the 17th generation to cultivate its soil. It is no longer a full-time occupation, though—he makes some extra money working as a carpenter, especially in the winter. There are still four generations of Sherwoods living on the farm—Tom's grandmother, his parents, himself and his wife Christine, and their two small children. The farm has been reduced in size over the years and is now only 36 acres. Tom's father, Schuyler Sherwood, grew mostly corn and tomatoes, with a few pumpkins for Halloween, but Tom has been branching out into eggplants, peppers, broccoli, onions, potatoes, carrots . . . "you name it," he says. They sell only what they themselves produce and usually have about 200 customers on Saturdays during the growing season.

Tom keeps bees to produce honey and to pollinate his crops, and he also has "a couple of hundred" hens for eggs. The hens live in a chicken house surrounded by a fenced-in area "where they go outside during the day and eat their insects and stuff like that. And of course they all fly over the fence . . . they come and go as they want, but as long as they go back in at night, it's OK." The hens have nesting boxes lined with pine shavings where they lay their eggs, which Schuyler collects and washes by hand. Tom doesn't debeak the chickens he raises himself, although some of those he buys have already been debeaked. "I never had a problem. I never had to debeak them." He's also never had a problem with his hens getting diseases from wild birds. At $2.25 a dozen, the eggs always sell out.

For Mary Ann, the freshness and flavor of Sherwood Farm's produce are big reasons for her to buy locally. Tom finds his customers are very discerning about taste, and if he grows a variety that doesn't have that home-grown flavor, they won't sell. He can grow varieties that supermarkets don't carry, bred for flavor rather than their ability to withstand transport and storage.

You don't have to live near a farm to buy from local farmers. On Saturdays, Manhattan's Union Square, traditionally the site of rallies to defend the rights of labor and other leftist causes, is now crowded with shoppers stocking up on fresh produce from farmers who have driven into the city early that morning with trucks full of fresh-picked produce. There were, at the last count, 47 greenmarkets spread across different New York neighborhoods, with 250,000 customers a week in peak season. New York City's Council on the Environment, which organizes the markets, allocates stalls only to regional farmers who produce the food they sell. No middlemen are permitted. Between them, the 175 farmers who regularly sell their goods in New York produce 120 different kinds of apples, a similar range of tomato varieties, and 350 kinds of peppers.[8]

The New York greenmarkets are part of a nationwide movement. St Paul, Minnesota, spent $2.2 million to renovate its downtown farmers' market, where more than 200 growers and 25,000 shoppers mingle on busy Saturdays. The market has established 13 branch locations around the city and extended operating hours. Kaiser Permanente, the health maintenance organization, is promoting good health by encouraging farmers' markets at its medical centers, where the customers are employees, patients, and local residents. The United States Department of Agriculture—a department widely criticized for being in the pocket of corporate agribusiness and hostile to small farmers—now hosts a farmers' market at its Washington headquarters and maintains a Web site that will tell you where your nearest farmers' market is. Similar information is available from nonprofit organizations that seek to promote local, sustainable agriculture, like FoodRoutes Network.[9]

In 2004 there were more than 3,700 farmers' markets in the United States—more than twice as many as there were in 1994. The number is growing at an accelerating rate, with nearly 600 new markets added over the previous two years, as compared to only 274 new markets in the two years from 2000 to 2002. More than 19,000 farmers told a U.S. Department of Agriculture (USDA) survey that they sold their products only at farmers' markets. All this is good news for those who enjoy the fresh, varied, and flavorsome goods sold at farmers' markets or local farms. The opportunity to chat to the people who produce your food, and to other shoppers, is also part of the attraction—a study has shown that people have ten times as many conversations at farmers' markets as they do in supermarkets.[10]

Nor is the renewed desire to buy locally produced food a specifically American phenomenon. In 1997 there was only one farmers' market in the whole of the U.K., but by 2002 there were around 450, and 70 percent of these describe themselves as thriving.[11] Farmers' markets were unknown in Australia until the 1990s—now they are in all the major cities. In Japan, householders concerned about pesticides and high supermarket prices organized themselves into a cooperative in 1965. That first cooperative rapidly spawned hundreds of others that today have 15 million members and buy billions of dollars worth of produce directly from Japanese farmers.[12] In Italy, Carlo Petrini, a journalist, founded the "Slow Food" movement to protect "endangered" local foods from being driven into extinction by global brands. The growth of the movement has been anything but slow, and it now boasts more than 80,000 members in more than 100 countries.

There is even a term for those who eat only local food: locavores. We first heard the term when a group of self-described culinary adventurers from the San Francisco Bay area decided that, for the month of August 2005, they would make an effort to eat only food grown within 100 miles of San Francisco. They set up a Web site, www.locavore.com, giving their reasons for doing so and advice about what to buy. August is a relatively easy month to eat locally in the northern hemisphere, because so many vegetables are ready to pick then. The Bay area locavores allow themselves a few imported treats: small quantities of chocolate, coconut milk, vanilla, black pepper, maple syrup, and Parmesan cheese, for example.

A Vancouver couple, Alisa Smith and James MacKinnon, set themselves a much more difficult task when they decided to live for an entire year on food grown within a 100-mile radius. Smith and MacKinnon started out as purists; no sugar, for example. Unable, at first, to find any locally-grown grains, they gave up bread, pasta, and rice. They made turnip "sandwiches" with slices of roasted turnip substituting for the bread, and ate a lot of potatoes. They wouldn't even eat locally produced, organic, free-range eggs because the hens were fed on grain imported from outside the region. Sometimes, walking into their "local" supermarket, they couldn't find a single thing to buy. Fortunately they eventually found a local wheat grower, and although they had to mill it themselves, they were soon joyfully eating pancakes and baking bread.

Is being a locavore just a challenging personal experiment in eating—something that might also be a lot of fun? Or is there a serious ethical argument for buying only locally produced food?

## ETHICAL ARGUMENTS FOR EATING LOCALLY

When Mary Ann and Jim were talking about buying locally-grown food, she mentioned that you are more likely to have some firsthand knowledge about the producer and its practices, and Jim said that, in addition to being fresher and riper, the food "hasn't cost an ecological fortune in fossil fuels coming across the continent." The "locavores" just mentioned give similar reasons. Jessica Prentice, one of the Bay area women who started the idea of eating locally for a month, urges people to eat within their own "foodshed," both to reduce fossil fuel use and because food loses its flavor when transported long distances. Alisa Smith cites a World-Watch Institute report as calculating that the average American meal uses up 17 times as much petroleum, and so is responsible for 17 times as much carbon dioxide emission, as a locally-produced meal would.

Several recent books make essentially the same case for buying locally. In *Eat Here: Reclaiming Homegrown Pleasures in a Global Supermarket,* Brian Halweil, a senior researcher at the Worldwatch Institute and a contributor to the report Smith mentions, asks: Isn't there something wrong with supermarkets in Des Moines selling apples from China when there are apple orchards in Iowa? Frances Moore Lappé is another big fan of local markets. Thirty years after her *Diet for a Small Planet* became a huge bestseller, she and her daughter Anna Lappé wrote *Hope's Edge: The Next Diet for a Small Planet*, championing the grassroots producers of local, community-based methods of growing, marketing, and eating food. Getting organic and local food, the Lappés write, is a decision "defining who we are."

Many advocacy groups, like FoodRoutes Network, are supporting the move to local food. Colleges, too, are joining the movement—more than 200 of them buy at least some food from local farmers. At Yale, after student groups began to push for local food, Yale president Robert Levin met with Alice Waters, owner of Berkeley's Chez Panisse restaurant and a renowned advocate of local, organic food. That led to the Yale Sustainable Food Project, a joint endeavor of students, faculty, and the university administration, together with the university dining services, to increase the amount of fresh local and organic food served in the university. So far, local organic food is available only at one of Yale's colleges—and students from other colleges who are not entitled to eat there are using fake ID cards to get in.[13]

The fact that local food is fresher and tastes better is not, in itself, an ethical reason for buying it. If I prefer convenience or low prices to freshness and good taste, that's my choice. Protecting my health and that of others for whom I buy food may be an ethical obligation, but local food is not necessarily healthier than other food. Local food is not always organically produced, and pesticide use may be less subject to checks when food is grown by small local farmers than by a corporate giant supplying Wal-Mart. In addition to freshness, taste, and health, FoodRoutes gives three additional reasons for buying locally that are broadly ethical. We'll consider them in turn.[14]

## 1. You'll Strengthen Your Local Economy

FoodRoutes says:

> Buying local food keeps your dollars circulating in your community. Getting to know the farmers who grow your food builds relationships based on understanding and trust, the foundation of strong communities.

Locavores.com takes much the same line, saying that "buying locally grown food keeps money within the community. This contributes to the health of all sectors of the local economy, increasing the local quality of life."

Is this an ethical reason for buying locally? The San Francisco Bay region is one of the wealthiest local economies on the entire planet. In developed countries, most local economies near major centers of population are doing very well, by global standards. If we have the choice of using our purchasing power in our local economy, or buying products imported, under fair terms of trade, from some of the world's poorer nations, is there any merit in keeping our money within our own community?

When we think ethically, we should put ourselves in the position of all those affected by our actions, no matter where they live. If farmers near San Francisco need extra income to send their children to good colleges, and farmers in developing nations need extra income in order to be able to afford basic health care or a few years of elementary school for their children, we will, other things being equal, do better to support the farmers in developing countries.

There are, of course, further questions to ask: is exporting food really going to help the poor in developing countries, or would they do better to become more self-sufficient rather than growing commodities for export? If you buy food from developing countries, how much of the money you spend will really go to the people who need it, rather than to transnational corporations? What are the environmental costs of transporting the food from the poorer countries? These questions all need to be investigated, and we will return to them in the following chapter. Until we do, we can't reach any conclusions about whether we should buy locally or from developing countries. Our point now is simply that "keep your dollars circulating in your own community" is not an ethical principle at all. To adhere to a principle of "buy locally," irrespective of the consequences for others, is a kind of community-based selfishness.

Building relationships based on understanding and trust can be a reason for buying locally. Mary Ann values the transparency that comes with personal knowledge and an interpersonal relationship with the person who grows her food. That is a sound reason for buying locally—as long as you are able to talk to the farmers and they are open to their customers visiting their farms and taking a look at what they do. But not everyone has time for that, and trust and understanding need not be exclusively local.

## 2. You'll Support Endangered Family Farms

FoodRoutes says:

> There's never been a more critical time to support your farming neighbors. With each local food purchase, you ensure that more of your money spent on food goes to the farmer.

The total number of farms fell sharply for most of the 20th century, and is still falling, if more slowly, now. The proportion of the population living on farms in the U.S. has fallen from nearly 40 percent in 1900 to less than 2 percent today. In the United States there are now only 1.2 million people whose primary occupation is operating a farm, making the U.S. a nation with more people in prison than in full-time farming.[15] A similar collapse in the number of farmers has occurred in all the developed countries and is now taking place in China as well.[16]

The three poorest counties in the United States are in Nebraska[17] Agriculturally-based counties generally have more people living in poverty and more low-income families than metropolitan counties. They also have a higher proportion of people under 18, but a lower proportion of those 18-44, suggesting that many young adults leave the farm and go to the cities. The upshot is that these counties also have twice the proportion of senior citizens as metropolitan counties.[18] Lacking children, schools are closing and once-viable towns have been abandoned.

Iowa once had a diversified agricultural base that supported thriving rural communities. In 1920, ten different commodities, including fruit and vegetables, were produced on more than half of Iowa's farms. But by 1997, that had fallen to two: corn and soybeans. With this increasing specialization, food processing plants closed. Most Iowa products now leave the state in unprocessed form. In 1920, about half of the apples eaten in Iowa were grown in Iowa; now the figure is down to 15 percent, and not many other locally consumed fruits and vegetables are grown there either.[19] Verlyn Klinkenborg, who grew up on an Iowa farm and writes on rural life for *The New York Times*, looks back on the Iowa of his boyhood and finds it difficult to imagine anywhere better to have been a child. But that idyllic place has been destroyed, he writes, by "the state's wholehearted, uncritical embrace of industrial agriculture, which has depopulated the countryside, destroyed the economic and social texture of small towns, and made certain that ordinary Iowans are defenseless against the pollution of factory farming." Klinkenborg's solution is to "try to reimagine the nature of farming."[20]

One way of reimagining farming is to bring together the people who grow the food and the people who eat it. When that happens, the farmers get to keep almost all of the dollars the consumers spend on food, instead of the roughly 20 percent—and falling—they otherwise receive.[21] Manufacturers, processors, advertisers, and retailers normally get all the rest. Transforming that situation could preserve family farms, keep people on the land, and revitalize rural communities, or at least those that are in reasonable driving distance of a metropolis. That would reduce the anguish of those people who are now watching their communities become modern ghost towns. Many farmers feel an irreparable loss at being unable to hold on to the family farm, a loss so deep it can lead to despair and even suicide. One survey has shown that five times as many U.S. farmers commit suicide as die from farm accidents.[22]

Rural depopulation is not itself a bad thing. If people in China prefer to move to the city to find employment rather than work long hours at hard physical labor on the land, it is good that they have this new option. Besides, some "traditional rural values" are better forgotten. Rural communities can be stultifyingly narrow and intolerant of diversity. Inevitably they produce fewer opportunities for people with unusual interests to meet others who share those interests. But some rural values are undeniably worth preserving. When people see themselves as custodians of a heritage they have received from their parents and will pass on to their children, they are more likely to cherish the land and farm it sustainably. If those people are replaced by large, corporate-owned farms with a focus on recouping the investment and making profits for a generation at most, we will all be worse off in the long run. So supporting endangered family farms can be an important value.

## 3. You'll Protect the Environment

According to Food Routes:

> Local food doesn't have to travel far. This reduces carbon dioxide emissions and packing materials. Buying local food also helps to make farming more profitable and selling farmland for development less attractive.

Reducing carbon dioxide emissions is an important ethical concern. Nine of the ten hottest years since reliable record-keeping began in 1861 have occurred since 1994. There is a broad consensus among scientists that human-generated greenhouse gas emissions are making a significant contribution to this pattern of global warming, and carbon dioxide is the most significant of these gases. The continuation of this pattern will mean more erratic rainfall patterns, with some arid regions turning into deserts; more forest fires; hurricanes hitting cities that at present are too far from the equator to be affected by them; tropical diseases spreading beyond their present zones; the extinction of species unable to adapt to warmer temperatures; retreating glaciers and melting polar ice-caps; and rising sea levels inundating coastal areas.[23] A far worse scenario cannot be ruled out: some scientists believe that the melting of the ice caps could release huge amounts of methane that accelerate further warming, forming a cloud layer so dense as to block out heat from the sun and cause the planet to go into a deep freeze that extinguishes life on earth.[24]

Mitigating global warming is therefore a major issue, and because what every nation does has an impact on every other nation, it is an ethical issue. No nation owns the atmosphere or has the right to use more of it than other nations. The United States, with less than five percent of the world's population, emits about a 25 percent of the world's greenhouse gases, more than any other country. In ethical terms, this means that the U.S. is currently using far more than its fair share of the capacity of the atmosphere to absorb our waste gases. That statement holds true on any plausible criterion of fairness. Does fairness consists in everyone having an equal share, the rule we usually use for slicing up a cake if everyone wants as much cake as they can get? By that rule, the U.S. takes about five times as much cake as it should, if we regard the current global level of emissions as sustainable. However, since virtually all experts agree that this level of pollution is too high, the U.S. is emitting more than five times the amount it should. Is the principle "the polluter pays" fair, as we usually think when a factory pollutes a river? Then the U.S. should be paying, as it is not only the biggest polluter now, it has been for the past century or more, and most of the gases emitted by the U.S. over that period are still up there in the atmosphere. Or does fairness require that the best-off should sacrifice more to help the worst-off, and those who have the greatest capacity to help should do the most? The U.S. is one of the richest nations in the world and has more capacity to help than any other nation.

When the United States refused to sign the Kyoto Protocol, it made other nations bear the burden of taking the first steps toward dealing with the problem of global warming.[25] (The Kyoto Protocol is not, in itself, enough. The cuts in emissions need to go much deeper, and developing nations like China and India will eventually need to be brought in; but it is, at least, a first step.) The United States, by continuing its high, and still increasing, level of emissions, is putting at risk the lives of tens of millions of peasant farmers whose land may turn into desert, or whose low-lying but fertile delta regions in Bangladesh and Egypt may be flooded by rising sea-levels.

The long distances that food travels in the United States is part of the high level of energy use in the U.S. food system as a whole. Food production, processing, manufacturing, distribution, and preparation consumes somewhere between 12 and 20 percent of the U.S. energy supply.[26] Per capita, the U.S. uses more energy for food production, processing, and distribution than Asia and Africa use for all activities combined.[27] Nevertheless, transportation of the food is, according to one study, responsible for only 11 percent of the total energy used in the food system, as compared with, for example, home preparation, which uses 26 percent, or processing, which consumes 29 percent of the total.[28] Nor is all transport equal in energy use. Transporting a given amount of food by plane uses the most energy per mile, almost twice as much as road freight and 20 times more energy than sending it by ship or rail.

While we agree that we have an ethical obligation to reduce carbon dioxide emissions and that transporting food long distances requires energy and produces carbon dioxide emissions, it does not follow that we can always reduce carbon dioxide emissions by buying locally produced food. At Sherwood Farms, Tom Sherwood has put in a hydroponic system to raise early tomatoes in a greenhouse. He hopes to get them a month earlier so he won't have to turn customers away in June when they come asking if the tomatoes are ready. He can also extend the tomato season into October. Much of the energy that ripens these tomatoes is solar heat trapped by the glass, but Tom has also put in an oil furnace that he estimates will cost him about $700–$800 dollars a year in extra fuel—at the prices current when we talked to him, that's 350 to 400 gallons of heating oil. The greenhouse will hold 330 plants, and at around 20 pounds per plant, should produce 6,600 pounds, or 3.3 tons, of tomatoes. To recover his energy costs, he'll sell them for $2.50 a pound, a 50 cent surcharge on what he charges for tomatoes from the field. His customers will be happy to pay that price, because Sherwood Farm's vine-ripened tomatoes taste a lot better than supermarket tomatoes trucked in from Florida or California, most of which are picked green and ripened with ethylene gas.

If Mary Ann is concerned about reducing the impact of her purchases on energy use and carbon dioxide emissions, she should compare the amount of oil Tom will burn to warm his greenhouses with the amount it would take to truck a load of tomatoes from, say, Florida, where they can be grown without artificial heat. We did a quick back-of-the-envelope calculation and came up with the conclusion that you could truck up from Florida the same quantity of early season tomatoes Tom was going to grow in his greenhouse for less than half the amount of fuel he was going to burn to produce them.[29] In other words Mary Ann would reduce her contribution to greenhouse gas emissions by avoiding her local farmer's greenhouse-grown early tomatoes and buying tomatoes from Florida. This outcome is especially striking, given that Tom is using heat only to combat late frosts and assist the sun to grow the tomatoes earlier than he otherwise could. The most profligate tomatoes of all, in terms of energy usage, are those grown in heated hothouses in northern countries like Canada and the Netherlands and then exported to the United States. If they travel from the Netherlands by air, that is worse still.

The conclusion of that rough calculation points in the same direction as more careful studies of the energy impact of similar choices in other countries. A British study carried out for the Department of Environment, Food and Rural Affairs showed that buying local tomatoes outside the usual outdoor season was responsible for three times the carbon dioxide emissions caused by growing the tomatoes in Spain and trucking them to Britain.[30] Of course, every situation is different, depending on the climate, the transport costs, the produce, the methods of production, how the greenhouse is heated, and, often, the season when the produce is bought. A Swedish study gave similar results to the British study for tomatoes, but showed that energy was saved when Swedes bought domestically-produced carrots rather than Italian ones, because even in Sweden, carrots don't need artificial heat.

## FLYING HIGH, SHIPPING LOW

The increasing amount of food being sent by air is a major problem, because air freight uses almost twice as much energy per ton/mile as road freight. Currently, about half of the freight sent by air travels in the hold of passenger flights when they have spare capacity, which is more efficient than sending it on freight-only aircraft, but the use of air freight is growing more rapidly than passenger travel, and so more freight-only aircraft are flying. It has been predicted that aviation will account for 15 percent of all greenhouse gas emissions by 2050. Although most of that will still be from personal travel, air freight will account for an increasing proportion of that very significant total, and by 2050 could make up nearly a third of the total commercial aviation fleet. Moreover, some experts believe that aviation makes a contribution to the greenhouse effect that goes beyond its energy use, because planes put particulates and water vapor into the upper atmosphere and create additional cloud cover. All of this has a heat-trapping effect that is difficult to quantify but could double that caused by carbon dioxide emissions alone.[31]

The environmental problems of air travel can create some ethical dilemmas. In Chapter 8, we described the genuinely free-range New Zealand eggs available in the western United States. Eggs are much lighter, on a calorie-to-weight basis, than tomatoes, but even so, it takes the energy equivalent of almost a gallon of diesel fuel to fly three dozen large eggs (weighing about 4 pounds, including the cartons), from Auckland to Los Angeles. Is it justifiable to use that amount of energy to give hens a better life? If no other humanely produced eggs are available, maybe we shouldn't be eating eggs at all.

If air freight is the most energy-extravagant way of moving food, sending it by sea or rail are the most economical ways. Rice is grown in California, under irrigation, but it takes a lot of energy to grow it there—about 15 to 25 times as much energy as it takes to grow rice by low-energy input methods in Bangladesh.[32] The energy used in shipping a ton of rice from Bangladesh to San Francisco is less than the difference between the amount of energy it takes to grow it in California and in Bangladesh, so if you live in San Francisco, you would save energy by buying rice that has traveled thousands of miles by sea, rather than locally-grown rice.

To put the energy involved in sea transport in perspective with other energy uses, taking the average car just five extra miles to visit a local farm or market will put as much carbon dioxide into the atmosphere as shipping 17 pounds of onions halfway around the world, from New Zealand to London.[33] That doesn't include the energy used to truck the onions to and from the docks in New Zealand and Britain, and it assumes that refrigeration was not required to store them, but it does show that proximity to the place of production is not necessarily a reliable guide to energy savings.

Other factors to take into account include the use of energy to sort, deliver, and store produce, especially if it has to be kept frozen, and to load and unload trucks, as well as tallying the efficiency of distribution to each store. A local farm like Sherwood may not incur those costs. But the situation is different for stores in town that sell local produce. Suppose an individual farmer has to take small quantities of produce to five local stores in five different nearby towns. Small vans use more fuel and emit more greenhouse gases, in terms of pounds per mile carried, than large trucks, so distribution to small stores may be less efficient, per pound, than an entire truckload of produce going to a large supermarket.

Then there is the customer's own energy use. Mary Ann lives less than five miles from Sherwood Farm and drives a Subaru Outback, so she doesn't use a lot of gas to go there. Still, her trip to Sherwood Farm to get those tastier tomatoes does use energy that she wouldn't have used if she had bought it all at Trader Joe's, because she goes there anyway for other items.

## CUTTING YOUR OWN ENERGY USE

To say that buying local food will reduce energy usage and hence carbon dioxide emissions is, at best, an oversimplification. The real story is much more complicated. People who do their shopping on foot, by bike, or by using public transport do best—but in developed societies today, the number who do that is decreasing. The British Department of Environment, Food and Rural Affairs study reported that the number of "food miles" traveled in urban areas in Britain has risen 27 percent since 1992, but this is largely because more people are using cars to do their weekly shopping, rather than walking to small local groceries. For many Americans, however, there simply is no choice—there are no local groceries in walking distance. Driving twenty miles in a big SUV to pick up eggs from a local farmer and then heading off in a different direction to get fresh local produce would almost certainly be less energy-efficient than buying everything at a single supermarket, even if the food has traveled further to get there.

Local foods, especially those bought at farmers' markets, are often unprocessed. Processes like freezing, dehydrating, and canning all use energy—but they may also reduce the amount of energy we use when we cook at home. The chicken processing industry, for example, argues that it is more energy efficient for consumers to buy pre-cooked chicken, which only needs to be reheated. That claim isn't entirely without merit, especially for people who use an electric oven—the least energy-efficient fuel for home cooking.[34] But there are likely to be other serious environmental and animal welfare problems with precooked chicken, as we have seen earlier in this book. If we really want to save energy, we should buy only fresh, unprocessed local food, grown outdoors, and eat it raw, or with minimal cooking.

Following that policy would mean restoring seasonality to fruits and vegetables. People living in northern regions of America or Europe would have to go without fresh tomatoes, lettuce, or strawberries for the winter and early spring. That might seem a hardship that it is unrealistic to expect modern affluent consumers to undergo. It would, however, have the compensating benefit of reviving the now-vanished excitement that used to greet the arrival of the first fresh fruits and vegetables of the season. That's no small gain. Carlo Petrini, founder of the Slow Food movement, argues that if not having some foods all year round is a constraint, it is much less of a constraint than "to be forced to eat standardized, tasteless industrial food products full of preservatives and artificial flavorings" and species of fruit and vegetables "with characteristics functional only to the food industry and not to the pleasure of food."[35]

Buying locally produced food is often the best ethical choice, but not *because* the food is locally produced. To reduce the amount of fossil fuel that is involved in producing our food, we should buy local food, if it has been grown with similar energy efficiency to food from somewhere else—but not if the local grower had to burn fossil fuel to provide heat, and not if there is a lot of extra driving involved in picking the food up, or getting it delivered.

"Buy locally *and* seasonally" is a better policy than simply "buy locally"—but it entails giving up a lot of fresh fruit and vegetables we have come to enjoy all year round. Supporting endangered family farms is a good reason for buying local, if our local family farms really are endangered and we are unable to buy from other, equally endangered family farms elsewhere. Transparency is often a good reason for buying local food, if you can visit the farm from which you buy. But sometimes the most environmentally friendly food is grown far away, under natural conditions more favorable to growing the food, and transport by sea is so efficient, in fossil fuel terms, that buying food from distant countries can contribute less to global warming than buying locally. San Francisco "locavores" would do better to buy imported rice from Bangladesh than to buy California rice. In Britain, about half of all organic food sold is imported. Buying imported organic food could be the more environmentally friendly choice, on a global level, than buying non-organic local food. Moreover, although it is certainly good to protect the environment and support local rural communities, we also have an obligation to support some of the word's poorest farmers, and under fair trading conditions, the best way to support them can be to buy the food they produce. We take up that topic in the next chapter.

# 11

# TRADE, FAIR TRADE, AND WORKERS' RIGHTS

In 1985, Larry and Sandra Jacobs, organic farmers from Pescadero, California, took a vacation in San Jose del Cabo, at the southern tip of Baja California, Mexico. Noticing that the land was fertile and well-watered and the winters mild, they had the idea that the area could grow organic vegetables in winter for export to the United States. They went to a local "ejido," or farming cooperative, and talked to them about organic agriculture. The farmers were poor, even by Mexican standards, left behind by the rising prices caused by the tourist boom along the Baja coast. They used synthetic fertilizers and pesticides, often without taking proper safety precautions for their own health. But they knew something about the kind of farming that the Jacobs were talking about, because it was what their grandparents had done. Ten farmers agreed to give it a try, working together and pooling their produce under the label Del Cabo. The Jacobs began by working alongside them, seeing how they were farming, and educating them about what they could and could not use, consistent with obtaining organic certification. When the certification was complete, the Jacobs found markets for the produce in the United States. Today the Del Cabo cooperative has about 300 small farms and earns $7 million a year from sales of organic vegetables. The farmers can send their children to school and feed their families more adequately than they could before. While many Mexican farmers are selling their land and going to live in the nation's crowded cities, the Del Cabo farmers are staying on the land.[1]

The Del Cabo cooperative stands as an ethical alternative to the idea that we should only buy locally produced food. For those of us living in the United States, Canada, Europe, Japan, Australia, New Zealand, and other developed countries, a decision to buy locally produced food is a decision *not* to buy food from countries that are significantly worse off than our own. If Americans were all "locavores," the small farmers of Del Cabo would still be living in poverty and almost certainly still using pesticides and chemical fertilizers—unless they had sold their land for tourist developments.

Is there an ethical obligation that points in the direction that the Jacobs took: to encourage trade, especially with those who are less well off than we are? We asked Mary Ann Masarech and Jim Motavalli whether, as a counterweight to their preference for buying local food, they considered other choices that might support poor farmers in developing countries. They replied that, with the exception of products like coffee and chocolate that have specific "fair trade" labels, they simply didn't have access to the information they would need to accomplish that. Mary Ann said "It's a great idea, I just don't know how easy it is to execute."

Let's start with some basic facts about global poverty. You may have heard the oft-quoted statistics that more than a billion people are currently living on about $US1 per day, and about 2.5 billion on under $US2 per day.[2] But these statistics, expressed like that, are misleading. As anyone who has visited poor countries knows, what you get when you exchange $US1 for the local currency buys a lot more than you could buy with $1 in the United States. So you might think that living on $US1 a day isn't so bad in some of those countries. But the statistics we've just referred to take that difference in purchasing power into account.

More accurately expressed, more than a billion people are currently living on less than what $US1 per day buys in America. So, at current exchange rates, they might be living on what 30 U.S. cents (or some other amount much less than a dollar) would buy if that sum were taken to an impoverished country and then converted to the local currency. That's a level of poverty that, for those of us living in the world's wealthy nations, is barely imaginable. It means that people cannot be sure that they will have enough to feed themselves or their families. If they or their children become ill—which is very likely, because they may not have safe drinking water—they cannot obtain even the most basic health care.

They cannot afford to send their children to school. This degree of poverty kills. Life expectancy in rich nations averages 77, whereas in sub-Saharan Africa it is 48. In rich countries, less than one child in a hundred dies before the age of five; in the poorest countries, one in five does. That means 30,000 young children are dying every day from preventable causes.[3] Three in every four of these extremely poor people live in rural areas in South and East Asia and sub-Saharan Africa.[4] Most of these rural households in poor countries depend on agriculture for whatever income they earn—it is the only option available.

Can we help some of these very poor rural people by buying food that they produce? In theory, that ought to be the case—if buying locally helps local farmers, it would seem that buying from developing countries should help farmers in those countries. Most economists believe that free trade tends to make people better off. Low labor costs give poor countries an important economic advantage over rich countries. The skills of the workforce, appropriate transportation and communications, and law and order are important factors as well, of course, but in the presence of free trade and the necessary infrastructure, some production will shift to countries with low labor costs, increasing the demand for labor in those countries. Once the pool of unemployed workers there has been soaked up, wages should rise. Thus, a free market should have the effect not only of making the world as a whole more prosperous, but more specifically, of assisting the poorest nations.

How well does this work for trade in agriculture? Ataman Aksoy, a consultant at the World Bank, and John Beghin, a professor in the Department of Economics and the Center for Agricultural and Rural Development at Iowa State University in Ames, edited a volume for the World Bank called *Global Agricultural Trade and Developing Countries* that addresses this question. Summarizing the findings of the fifteen studies of agricultural trade included in the book, they conclude that a reformed and more open system for trade in agricultural commodities would, overall, "reduce rural poverty in developing economies, both because in the aggregate they have a strong comparative advantage in agriculture and because the agricultural sector is important for income generation in these countries."[5]

Although non-government aid agencies are often at odds with the World Bank, in this case Oxfam International, the leader of the international campaign to make the rules of trade fairer to developing countries, agrees on the significance of exports to developing countries. In a 2002 report that was part of its campaign for fair trade, Oxfam said: "History makes a mockery of the claim that trade cannot work for the poor. Participation in world trade has figured prominently in many of the most successful cases of poverty reduction—and, compared with aid, it has far more potential to benefit the poor." The Oxfam report points out that if sub-Saharan Africa could increase its share of world exports by just one percentage point, that would generate new foreign exchange earnings equal in value to 20 percent of current total income in the region. The gains from a one-percentage-point increase in its share of world exports would amount to five times the aid that sub-Saharan Africa receives.

Charles Walaga, the Ugandan member of the International Federation of Organic Agriculture Movements' Development Forum, is an enthusiastic supporter of the development of an African export industry in organic produce, because he believes it will benefit small Ugandan farmers. He points out that in 1986, when Uganda had few exports and most smallholders were subsistence farmers, 56 percent of the population lived in poverty. By 2000, when economic liberalization had given Ugandan farmers greater incentives to sell to exporters, poverty had fallen to 35 percent.[6]

Some people question the assumption that people will be better off growing crops for export rather than using their land to grow food for themselves and their families. They wonder why countries in which people do not get enough to eat still export food, suggesting that if people are poor, they should keep all the food they grow. But except for people who are literally starving, food is not everything. In some regions, agriculture can provide the basics that allow families to become self-sufficient, but elsewhere they may be able to grow only a single staple crop that does not provide everything they need even to be well-nourished, let alone enabling the poor to raise their living standards above bare subsistence

With regard to Uganda, Walaga notes that about 80 percent of the population lives in rural areas, and the overwhelming majority of them are small holder farmers who work their own land and grow their own food. As a result there is not much of a local market for agricultural products, and farmers have little opportunity to earn income from selling their produce locally. So, Walaga asks, "Where would these small holder farmers get money for medical care, education, and clothing if they did not produce for the export market?"[7]

It is a mistake to think of growing food for oneself or for export as an either/or choice. Small farmers generally grow some crops for their own use and others for market. Often the export earnings enable them to be more successful in growing food for themselves. Consider Uganda, where earnings from coffee enabled farmers to diversify, buying pigs and goats for their own use, as well as learning to grow vegetables for export. Poverty has fallen ten times as fast among coffee producers as among those who grow only staple food crops. Alice Lukoba, a widow farming land about 40 miles north of Kampala, the Ugandan capital, told Oxfam's researchers: "Life is always hard for us farmers, especially now that our coffee prices are bad. But no farmer here will tell you that life today is harder than it was before. We were given an opportunity to get something out of our coffee—and we took it."[8] It is not only the farmers themselves who benefit from export earnings, either. When their income increases, they create demand for local services, including farm labor, construction, and clothes, and these flow-on effects benefit even those who have no land at all.

We wondered what Brian Halweil, the local food advocate and author of *Eat Here*, would have to say about this argument. In his response, he agreed that if Americans were to buy exclusively local fruits and vegetables, "Mexican garlic growers and Chilean berry growers would lose a big market." But he added that his impression was that "of all the money spent on food exports, very little gets back to the actual grower and farm community. Most is gobbled by food traders, brokers, and shippers . . . Personally, I think that First Worlders can best reduce poverty in their own backyard—among struggling rural populations and the burgeoning class of inner-city food producers—by eating local." (By "inner-city food producers," Halweil means community gardens and urban farms, which he grants are not yet producing on a significant scale, although he regards them as having great educational value in showing what can be done to produce healthy food.)[9]

Halweil is right to point out that when food passes through many intermediaries, often only a small fraction of what we pay for it gets back to the original producer. Researchers at the Institute of Development Studies at the University of Sussex found that Kenyan fresh fruit and vegetable producers received 14 percent of the retail price of their produce, while Zimbabwean growers of snow peas and South African peach growers retained about 12 percent. In the case of bananas, another study found that about 12 percent stayed in the producing country, but the plantation workers themselves received only 2 cents from every dollar the consumer paid.[10] Even in that last case, however, the small percentage that returns to the worker is insufficient evidence to prove that people in rich countries can do more to relieve poverty by buying locally.

In our view, Halweil has not sufficiently taken into account the staggering differences in income between the world's poorest farmers and America's poor rural farmers. As we have seen, some of the world's poorest farmers struggle to support their families on as little as—in exchange rate values—30 U.S. cents per day, or $110 a year. In addition, America offers domestic safety nets. Farmers can get food stamps; public schools are available for their children, and if they need health care they may be eligible for Medicaid or, after age 65, Medicare There are no food stamps in the world's least developed countries, schools are not always free, and the average *annual* per capita expenditure on health care in some of these countries is $US10.

Suppose that we pay a dollar for beans grown in a developing country. Out of that dollar, after everyone along the supply chain has taken their cut, perhaps the worker who grew them, like the banana plantation workers mentioned above, receives just two cents. If that worker is living on $110 a year, that two cents is almost one five-thousandth of his or her annual income. That may not sound like much, but then, we've only spent a dollar. If a thousand others also spend a dollar on those products, now we're talking about a nearly 20 percent increase in the farmer's income. That will make a bigger difference to the worker than the entire $1,000 would to an American farmer. When you are *very* poor, a small dollar increase in your income does more to improve your well-being than a far larger dollar increase does when you are much better off.

That leads us to a conclusion that may seem surprising: if you have a dollar to spend on beans and you can choose between buying locally grown beans at a farmers' market or beans grown by a poor farmer in Kenya—even if the local farmer would get to keep the entire dollar and the Kenyan farmer would get only two cents from your dollar—you will do more to relieve poverty by buying the Kenyan beans. This example is imaginary, but it illustrates how easily growth in agricultural exports can have an impact on rural poverty in developing nations.

Some advocates of organic agriculture don't like the idea of importing organic food. But Charles Walaga believes that the trend toward global trade in organic food is one of the forms of globalization from which Africa stands to benefit most. It will encourage African smallholders to farm sustainably and will make them independent of those who provide agrochemicals. Walaga asserts that organic agriculture is more in keeping with African culture than industrial agriculture, because "agriculture in Africa is not merely a business, as industrial agriculture is, but is also the culture of the people." He also argues that providing more income to small farmers in Africa will, in the long run, have environmental benefits. If farmers cannot educate their children, the children will have no alternative but to become small farmers themselves. But with a growing population, that will mean more forest clearing and more degradation of marginal lands.[11] It's also worth remembering that education has repeatedly been shown to lead to smaller families and thus to slow population growth, the root cause of so much environmental damage.

There is, in the end, an irresolvable tension in the idea of growing crops with techniques that take good care of the environment and then using jet fuel to send them half way around the world to environmentally-aware consumers. How are we to weigh environmental costs against social gains? That question raises deep philosophical and factual issues that we cannot resolve here. However, there is a strong case for buying from the least developed countries, at least when it comes to products transported by ship rather than plane and when a significant proportion of the purchase price is likely to end up in the hands of low-income farmers.

## FAIR TRADE

As we have seen, Jim and Mary Ann prefer to buy fair trade coffee and chocolate. But the term "fair trade" needs some explanation before we can assess whether we should be seeking out fair trade products.

The fair trade labeling movement began in the 1980s when coffee prices were falling—a fall from which, despite a recent increase, they have never fully recovered. Most of the world's coffee is grown by 25 million small farmers in developing countries. They usually have no more than 25 acres of coffee trees and depend on their crop for their livelihood. The fall in coffee prices left some of these small coffee growers without enough income to survive.

A priest working with Mexican coffee growers and a Dutch non-government organization had the idea of a special brand of coffee that gave a fair return to the growers. Introduced in the Netherlands in 1988, Max Havelaar Coffee—named after a fictional Dutch character who opposed the exploitation of coffee pickers in the Dutch colonies—soon captured 3 percent of the market. Many Europeans liked the idea of entering into a direct relationship with the small farmers, thus cutting out what they considered to be the excessive profits made by the multinational corporations. Soon other "fair trade" labels were appearing on tea and chocolate.

But the differing standards and labels caused confusion. In 1997, groups from 17 countries formed Fairtrade Labeling Organization International, or FLO, based in Germany, to harmonize international fair trade standards for certification and labeling. That gave the fair trade movement a further boost. Cafédirect, founded by the aid agency Oxfam and other non-government organizations, grew to become the sixth-biggest coffee brand in Britain, and national chains like Costa Coffee and Pret a Manger began selling fair trade coffee.

To receive the fair trade label, a product has to comply with standards all along the supply chain—including producers, traders, processors, and wholesalers. Small farmers are required to be organized in cooperatives or other groups that allow democratic participation. Plantations and factories can use the fair trade label if employers pay their workers decent wages, comply with health, safety and environmental standards, allow workers to join unions or other forms of associations, provide good housing if workers are not living at home, and do not use child labor or forced labor. Companies must ensure that the additional revenue generated by receiving fair trade prices for the products is used for the benefit of the workers. Beyond the minimum standards that producers must meet in order to be certified, FLO also requires continuous progress in worker conditions, product quality, and environmental sustainability. Traders seeking fair trade certification must pay producers a price that covers the costs of sustainable production and gives a return producers can live on, plus a premium that the producers can invest in development. For coffee, for example, the minimum price is $1.26 per pound, no matter how low the normal market price may fall. If the market price rises above that figure, the fair trade price will increase so that it remains 5 cents per pound higher. Moreover, this amount must go directly to the growers, without deductions from middlemen. If the producers need partial payment in advance, to cover costs, the traders must be willing to pay it, and they must be prepared to sign contracts that allow for long-term planning and sustainable production practices. To date, fair trade certification has been granted to coffee, tea, cocoa, rice, juices, sugar, spices, herbs, cotton, flowers, honey, wine, rum, bananas, mangoes, pineapple, and a variety of other fresh and dried fruit.[12]

According to Transfair USA, the American arm of the fair trade movement, fair trade coffee is now the fastest growing section of the specialty coffee market in the United States. It is available at major chains like Starbucks and Dunkin' Donuts, McDonald's, Costco, and Sam's Club. U.S. sales of fair trade coffee have generated an additional $31 million that has been passed back to the growers. Some of that money enabled a cooperative in Papua New Guinea to invest in a medical team to provide health services that were otherwise unavailable in the isolated rural area where they lived. The La Voz cooperative, in the highlands of Guatemala, used its proceeds from fair trade sales to help its indigenous Mayan members to send some of their children to college for the first time. A cocoa cooperative in the Dominican Republic invested its earnings to improve the quality of its cocoa and convert to certified organic production, which improved its position in the export market. Fair trading can, Transfair USA claims, even help reduce the illicit drug trade, by providing economically viable alternatives to the growing of coca and opium poppies, the raw materials for cocaine and heroin.[13]

The fair trade movement has helped to focus attention on equity and sustainability issues even when food does not actually have fair trade certification. Starbucks, for example, in addition to selling some coffee that is fair trade certified, has also developed its own program, called Coffee and Farmer Equity Practices (CAFE Practices), to assess the "economic, social, and environmental conditions" under which coffee is grown and give preference to farms that rank the highest.[14]

Under the influence of a growing sensibility that corporations should act in a manner that is socially responsible rather than one that aims only at maximizing short-term profit, some corporations are bringing their operations into compliance with a new international standard established by Social Accountability International. This New York-based non-governmental organization has an advisory board that includes representatives of Amnesty International, international trade unions, Care International, and such major corporations as Chiquita, Dole, Eileen Fisher, and Coop Italia. This unusually diverse board has been able to agree on a standard known as Social Accountability 8000, or SA8000, which focuses on the rights of workers, as laid down in declarations such as the International Labor Organization's Declaration of Fundamental Principles and Rights at Work. To gain SA8000 certification, a facility must pass an independent audit that assesses its performance in nine areas:

- **Child Labor**—with some limited exceptions, all workers must be at least 15 years old.
- **Forced Labor**—there must be no forced labor, which also means no prison labor and no debt bondage (forced labor to repay real or false debts).
- **Health and Safety**—the employer must provide a safe and healthy work environment; including access to bathrooms and safe drinking water.
- **Freedom of Association and Right to Collective Bargaining**—the employer must respect the workers' rights to form and join trade unions and bargain collectively.
- **Discrimination**—there can be no discrimination based on race, caste, origin, religion, disability, gender, sexual orientation, union or political affiliation, or age, and no sexual harassment.
- **Discipline**—no corporal punishment, mental or physical coercion, or verbal abuse may be used.
- **Working Hours**—must comply with the applicable law but, in any event, cannot exceed 48 hours per week and must allow for at least one day off per week. Overtime must be voluntary, paid at a premium rate and not to exceed 12 hours per week on a regular basis.
- **Wages**—must meet the legal and industry standards and be sufficient to meet the basic need of workers and their families.
- **Management Systems**—facilities seeking to gain and maintain certification must go beyond simple compliance to integrate the standard into their management systems and practices.[15]

For decades, the multinational corporations involved in growing bananas in Latin America and the Caribbean for export to the United States and other developed countries have been notorious for requiring their plantation workers to labor for long hours at low pay. They have used toxic chemicals hazardous to the health of the workers and have used intimidation or outright violence to make sure that workers do not unionize. It is significant, therefore, that the first major agricultural operation to gain SA8000 certification was Chiquita, one of the biggest international producers and suppliers of bananas and other fruit.

In 2005, Chiquita announced that it had achieved certification for all of the farms it owned in Latin America, covering more than 14,000 employees and 37,000 acres of farms. These farms also comply with the environmental standards of the Rainforest Alliance. Since Chiquita also buys bananas from other producers, this does not mean that all bananas labeled "Chiquita" have been grown on farms that comply with the SA8000 standard. But by March 2005, according to Michael Mitchell, Chiquita's director of corporate communications, about 43 percent of Chiquita bananas came from farms that met the standard, and Chiquita was actively working with its suppliers to increase that figure. Chiquita does not see compliance with SA8000 as a direct marketing device, like the "organic" or "fair trade" labels, but rather as, in Mitchell's words, "a way to improve our social and environmental performance and ensure that employees throughout the company understand our Core Values and the standards we use to measure our performance."[16] But as consumers become more ethically aware, meeting a measurable standard of decent treatment for workers can be a valuable defense against charges of exploitation.

Although the case for buying fair trade products is broadly similar for the entire range of products available, including coffee, tea, cocoa, fruit, sugar, in the case of cocoa—the basic ingredient in chocolate—there is an additional factor that makes the ethical case for buying only fair trade brands especially strong. In 2001, a British television investigation claimed that many child workers on West African cocoa plantations were, effectively, slaves. As the evidence of trafficking in children for labor accumulated, U.S. Senator Tom Harkin, a Democrat from Iowa, proposed legislation to ensure that chocolate products sold in the U.S.

were made from cocoa grown in compliance with the International Labor Organization's standards on child labor. When the industry announced that it would develop a voluntary system that would achieve the same result, Harkin agreed to shelve his proposals.

To gather information, the industry commissioned the International Institute of Tropical Agriculture to carry out a survey into child labor in Ivory Coast, Cameroon, Ghana, and Nigeria, which together grow 70 percent of the world's cocoa. The survey found that 64 percent of children working on cocoa farms were under 14 and that as many as 153,000 of them were engaged in the potentially hazardous task of applying pesticides. Even larger numbers of children used machetes or carried heavy loads. The report does not use the term "slave," but it does state that nearly 12,000 children working on cocoa farms had no family ties to the cocoa farmer or to local farm workers, and 2,500 had been recruited through intermediaries for cocoa farming. Approximately one-third of school-age children living in cocoa-producing households in Ivory Coast had never been to school.[17]

Despite these disturbing findings and the chocolate industry's assurances, at the time of this writing the promised voluntary system is still not in place. When the industry told Senator Harkin that it would not meet a July, 2005, deadline for certification that it be in compliance with international standards on the use of child labor, the Senator announced that he was going to buy his wife flowers for Valentine's Day, instead of chocolate.[18]

That may have been an effective publicity move, but in the intervening years the availability of fair trade chocolate had increased, and Harkin would have been on safe ethical ground if he had bought his wife Divine chocolate, made by the Day Chocolate Company. Day Chocolate is owned by Kuapa Kokoo, a democratically run cooperative in Ghana that now has 35,000 members from 460 villages. The standards for fair trade chocolate do not permit forced or bonded labor, nor may children work in circumstances that could jeopardize their education, or lead to them performing hazardous tasks.[19] In addition to receiving a price that is higher than the usual market price for its fairly traded cocoa, the cooperative receives a "social premium" of an additional ten percent that goes to fund bore holes for wells, health centers, and schools. Kuapa Kokoo runs special programs to educate women and encourage them to earn their own money, and 30 percent of its members are now female.

Fair trade schemes have generated hundreds of millions of additional dollars for small producers in developing countries and assisted those who produce our food and drink to belong to democratically-run cooperatives or work under conditions that allow them freedom of association and protection from sexual harassment. So buying fair trade products seems obviously more ethical, for those who can afford it.

Yet not everyone supports fair trade schemes. Brink Lindsey, director of the pro-market Cato Institute's Center for Trade Policy Studies, has argued that the fair trade campaign is a "well-meaning dead end." With some justification, he argues that the real cause of the fall in coffee prices was not the profiteering of multinationals, but big increases in coffee production in Brazil and Vietnam, combined with new techniques that make it possible to grow coffee with less labor and hence more cheaply. So if we want to assist coffee growers, Lindsey urges, we should encourage them either to abandon coffee and produce more profitable crops—and here he rightly points to the rich nations' trade barriers and subsidies as obstacles that need to be dismantled—or to move into higher-value products, like specialty coffees, that bring higher prices.[20]

What is curious about Lindsey's argument, however, is that the fair trade campaign can be seen as doing just what he himself recommends—encouraging coffee farmers to produce a specialty coffee that brings a higher price. He thinks that encouraging growers to produce under fair trade conditions distorts the market, but that encouraging growers to produce coffee that consumers will buy because they like its taste is just the normal operation of the market. But as long as consumers freely buy a product—whatever the reason they have for buying it—there are no grounds for saying that the market has been distorted. Pro-market economists accept consumers paying $48 a pound for Jamaica Blue Mountain coffee (the coffee that James Bond prefers), regardless of whether it tastes better than cheaper coffees. They don't object to corporations that blatantly use snob appeal to promote their products. So why criticize the decision of some consumers to pay $12 for a pound of coffee that they know has been grown without toxic chemicals, under shade trees that help birds to survive, by farmers who can now afford to feed and educate their children?

There is no economic—let alone ethical—reason why people's purchases must be driven exclusively by self-interest rather than a desire to help others. There is, however, some evidence that studying economics can make people more self-interested. In a delightfully simple piece of research, Robert Frank, a professor of economics at Cornell University, asked American university students at the beginning and end of a semester whether they would return a lost envelope with $100 in it. Students who had taken an economics course during the semester showed a greater shift away from returning the envelope than students who had taken an astronomy course.[21] Selfishness, however, is not a universal norm. Hundreds of experiments carried out in many different countries show that people do care about fairness and will give up some economic advantage because they refuse to accept an outcome that they see as unfair.[22]

Economists might reply that if you want to help people feed and educate their children, you can do so by paying $10 a pound for coffee that is not fair trade but tastes the same as the $12 per pound coffee that is, and then giving the $2 you have saved to an aid agency that provides food and education for poor children. That's a possible strategy, and we wouldn't say that anyone who does exactly that is acting unethically. But the premium prices for fair trade coffee are not charity. The growers know that they have to produce the coffee to receive their income. They also know that they have to produce a product that consumers like, both for its taste and for the way it is grown. If their product sells well, they can take pride in having produced something that is sought after around the world. From the growers' perspective, receiving a premium by selling a fair trade product is preferable to receiving a charitable handout that they would get whether they worked or not, and irrespective of the quality of what they produce.

Some consumers may buy the fair trade product simply because they feel they can trust the quality and ingredients of the product to a greater extent than one produced under greater pressure to keep prices low. Other consumers may prefer buying fair trade to giving to a charity because they fear that a charitable donation would involve giving their money to people who are unwilling to work. (In fact, aid agencies are unlikely to give money to someone who is too lazy to work—but some believe they do.) Paying more for a fair trade label is no more "anti-market" than paying more for a Gucci label, and it reflects better ethical priorities. Fair trade is not a government subsidy. Subsidies distort markets because the producer gets a government payment on top of whatever they receive from consumers. By contrast, the success of fair trade coffee is fully dependent on market demand.

Of course, it is true that fair trade coffee will not raise the returns to all coffee growers, but it is a mistake to think that because a proposal cannot solve a very big problem it cannot do any good at all. If more people buy fair trade coffee, more small farmers can make a decent living from growing coffee. For that reason, if you buy coffee, it is better to buy fair trade coffee. The same is true of chocolate, tea, sugar, bananas, and other products—although in the case of bananas, buying Chiquita bananas is the next best option, since there is a good chance that they will have been produced on farms that provide basic rights for workers.

# 12

# EATING OUT AND EATING IN, ETHICALLY

When Jim and Mary Ann eat out, they try to find a place that can meet their ethical standards. The family often looks for restaurants that serve vegetarian and organic food, but they aren't always easy to find, and they are often more expensive. Jim, of course, won't order meat, and tries to avoid fish from stocks that are endangered. But he admits that he sometimes has to loosen his standards, because when he asks where the fish comes from, the result is often "complete bafflement" on the part of the waiter. "If we found a place that gave you a lot of good information about where the food came from," he said, "I would tend to go back there. I would want to support their doing that." "If it was within your budget," Mary Ann adds.

In almost any eating situation, you can make more ethical and less ethical choices. If you find yourself in a restaurant that doesn't offer any guidance on how the animals from which its meats came were raised or whether its seafood came from a sustainable fishery, asking the server is always a good idea. Waiters baffled by such questions will soon learn that they need to find the answers. If you can't get a satisfactory answer, you may want to leave. Some social situations make that difficult or nearly impossible—if you are the only member of your family who cares about such things, you can't walk out on your mother's birthday dinner. But even if you are not going to leave and will end up eating the least problematic option on a generally unethical menu, asking questions can be part of a process of producing change. When the server has to go to the kitchen to make enquiries, you know that the message is getting through to someone else.

In this chapter we look at three models of ethical business in the food industry. What compromises have to be made to be ethical in business? We don't say that the three businesses we describe are doing everything right, but we do think they provide worthwhile examples of how it is possible to combine business success with ethical standards.

## A RESTAURANT: THE WHITE DOG CAFÉ

The White Dog Café in Philadelphia is a place for good food—with layers of meaning in the word *good*. "Capitalism for the common good" is how its owner, Judy Wicks, thinks of it. "The basic purpose of a business is to serve," she says. "White Dog's mission is to serve in four areas: serve customers, serve our employees, serve our community, and serve nature."

The customers can, if they wish, start at a lively bar stocked with single malt scotches, locally distilled rum, grappa, a local beer ("White Dog Leg Lifter Lager"), and cocktails with names like "Shar Pei" and "Pink Poodle." (Judy Wicks loves dogs and has a wonderful sense of humor.) We ate in a large airy room that looks out on sycamore trees planted by Judy Wicks years ago.The menu features, among other things, "Winter Harbor's Wild Caught Nova Smoked Salmon over Warm Sweet Pea Pancakes," "Pan Seared Corn Crusted Local Seitan with Balsamic Stewed Onions," and other vegetarian and meat dishes. We had the seitan, made by Philadelphia's own Ray Reichel, and found it tender and juicy, with a robust flavor that would appeal to meat-eaters too. In addition to the onions, we tried roasted Lancaster County sweet potatoes and garlic-sautéed spinach and zucchini. For dessert, there is "Grapefruit Pistachio Tart (with orange syrup and mascarpone mousse)," "Dark Chocolate Chipotle Pot De Crème," and a lot more.

The White Dog serves its employees by paying all of them a living wage—an amount based on the cost of living in the city, supplemented by an offer of $125 per month towards their health care insurance. On top of that, they divide up 10 percent of the profits of the business. Community causes share another 20 percent of the profits via the White Dog Foundation, which assists local businesses, farmers, and social justice groups. Judy evidently enjoys redistributing the wealth of her $5 million-a-year business according to ethical principles—she laughs heartily as she tells us about it. Her adjoining Black Cat retail store is stocked with items produced by fair trade and sustainable methods, many handcrafted by indigenous people. All the restaurant's electricity payments are allocated to wind power generation, which comes from windmills financed by a local affiliate of the Business Alliance for Local Living Economy, which Judy co-founded. She also started a project to bring locally grown food to the Philadelphia marketplace and thus support humane and sustainable farmers in the region.

Judy comes from a family of conservative, small-town Republicans. In 1983, she started running a coffee and muffin take-out shop from the first floor of her house on a block of then run-down Philadelphia Victorian brownstone townhouses. Gradually she added home-made bread and soup, then lunch, and then—using a charcoal grill in her backyard—dinner. But the political and ethical orientation of her business changed because of a trip she made to Nicaragua during the Reagan years. "I was crying in the airport and I thought, 'Why am I so upset?' I realized I was crying for my country. I had been so patriotic all my life. I was just heartbroken over what we were doing there." Judy began to see her business as a possible vehicle for social justice. She started by making the White Dog a sort of salon, a place to meet and discuss current events and social issues. She held talks and slide shows about her trips to other countries. She brought in speakers who were expert on the environment, education, the arts, and other concerns. Today, White Dog Café features speakers on Monday nights and attracts crowds with the food, the service, and the ambience.

Since Judy makes a point of running her restaurant along ethical lines, she is often asked why she hasn't made the restaurant vegetarian. She explains that she was a vegetarian for years and had a big conflict about it when she built up her business. She decided that a vegetarian restaurant would be "preaching to the choir." She wanted to bring in ordinary people, not already converted to her way of thinking. She admits, though, that at first she was naïve about the conditions in which animals are raised. When she heard about factory farming, it was the conditions for pigs that especially bothered her. "I thought, 'Oh my gosh, I can't go vegetarian because that could put me out of business,' but I also can't contribute to that system. So I went to the kitchen and said, 'We're going to find a source for humanely raised pigs.'"

Judy turned to the Lancaster County, Pennsylvania, Amish farmer who was supplying her pasture-raised chickens and eggs from hens not confined to cages. She had worked with him for many years and she knew that he understood her values. At the time, Judy did not know much about evolving standards for humane pig farming. "I just relied on the farmer. The main thing was, I wanted to make sure that the sows were not kept in crates." She found that the sows were given deep straw bedding, which passes the Animal Welfare Institute's guidelines. But after Judy educated herself about humane pig farming systems, she decided that pigs should have access to pasture. Now she buys most of her pork from Meadow Run Farm, another Lancaster County, Pennsylvania, farm, but occasionally buys from Niman Ranch when she needs extra bacon or pork chops.

White Dog Café puts its ideals and standards right up front. Its brochure, "Where does our food come from?" explains that not only do seasonal, local, and ecologically sound foods and ingredients taste better, but that "eating is a political act" and likens food choices to voting for a better farm and food policy. White Dog selects meat and poultry only from "humane agriculture" that "recognizes an animal's right to clean water, fresh air and sunshine, a healthy diet free of chemicals and hormones, appropriate shelter, freedom to move out of doors, and the opportunity to socialize and live in family groups." There are also standards for humane handling, transportation, and slaughter. Eggs, cheese, and other dairy products must come from local, organic, and humane sources "whenever possible." Seafood must come from sustainable fisheries.

Judy buys organic ingredients "as much as possible," she says. She believes that it is more important to buy local than it is to buy organic. She feels that the personal relationship of trust and support of the local food system "is the number one thing. And once you're supporting these smaller farmers, you can move them toward organic, you can increase the demand for that." She explains that a lot of the farmers she buys from are following organic standards but are not certified because of the expensive process involved. "When you work with people you know and you're there on their farms and you trust them, the certification is not so important."

Coffee, obviously, can't be grown locally, but Judy does have a personal relationship with her Mexican coffee growers. She often travels to

Chiapas, where the indigenous Zapatistas have been involved in a long struggling for land and basic rights. She has provided financing to help them grow coffee organically and sell it to the United States. Although it is not fair trade certified, she says she knows firsthand that she is paying a fair price. The tea she serves is organic and fair trade certified, and she is currently involved in ensuring that all of the chocolate used in their desserts is also fair trade. "That's in process," she tells us. "Our dark chocolate was organic and Fair Trade, but our milk chocolate was not, so we're just now switching over all of our chocolate. But it is much more expensive. Something like double. We're working on how much we're going to have to charge for our chocolate pie."

That leads us to ask Judy about the cost of dining at the White Dog. "We are more expensive than some restaurants," she tells us, "but we're not as expensive as others that don't buy fair trade, or organic, or local, seasonal foods." She tells us that she's gone to fancy restaurants and discovered that their food comes from the corporate system. They are making larger profits per meal served than she is, but she feels that her prices reflect more of the true cost of farming and food production. "I feel that the good that you do comes back to you. Our customers appreciate the fact that we buy direct from farmers, and that increases our business."

## A FAST FOOD CHAIN: CHIPOTLE

Most of the corporations that supplied the food our families ate ignored our requests for information about their production methods. But one morning the phone rang and a voice said: "Hello, Jim? Steve Ells, from Chipotle . . . Chipotle Mexican Grill. I'm intrigued by your project. I don't know if you know anything about Chipotle . . . "

Steve Ells is the founder and CEO of Chipotle. He opened the first Chipotle Mexican Grill restaurant in the Denver area in 1993, based on the idea of using fresh ingredients to provide great tasting burritos and tacos, served quickly. The concept brought in customers. Soon there was a second location, then a third. Ells went about looking for capital to expand the chain, but he found investors wary, until he made a deal with McDonald's. They brought the capital, networks, and human resources that helped Chipotle grow and eventually took a majority stake in the chain. Of the relationship, Chipotle's Chris Arnold says, "They were impressed from the first visit. They've recognized that the way we do things is very different and they've been largely hands-off, leaving the running of the business to us." (In more staid corporations, Arnold might be called Vice-President for Public Relations, but at Chipotle, it's Director, Hoopla, Hype & Ballyhoo.)

Then in 2000, Ells started to take Chipotle in a new direction. "I had the idea," he told us, "that just because it's fresh, doesn't mean it's good enough." He wrote a manifesto called "Food with Integrity," which "means working back along the food chain. It means going beyond distributors to discover how the vegetables are grown, how the pigs and chickens are raised, where the best spices come from." That's why Ells was excited enough about our book to call us personally. He wasn't surprised to learn that many companies had not wanted to tell us about their suppliers. "Of course they don't," he said, "it's not something that they would be proud of."

In 2000, after reading about Niman Ranch in a food magazine, Ells sought out Bill Niman and asked for a sample. At the time, he had about 60 restaurants. For the pork in his burritos and carnitas, he started buying Niman Ranch's pork, raised on farms like those owned by Mike Jones and Tim Holmes. Customers didn't flinch at paying an extra dollar for Niman's pork. Chipotle's purchases were useful for Niman, because they didn't need the most popular cuts—such as chops and loins—that were in demand from other restaurants and specialty stores. Chipotle also uses other pork suppliers, but all of them must allow their pigs adequate space to move around, either by access to the outdoors or in pens deeply bedded with straw.

Chipotle is known for using irreverent humor in its advertising—check out the company's Web site, www.chipotle.com, and you'll see what we mean. Ells does not say much about the pig farms he buys from on the menu, but he says plenty in his advertisements and restaurant posters. One advertisement says, "Eat a burrito. Help a family farm." It shows a photo of one of Chipotle's pork suppliers—three generations of the Willis family in front of their barn. Another shows Iowa farmer Duane Dorenkamp standing by his barn and says, "We know exactly where our pork comes from. Duane." (Most fast-food chains have no idea where their meat comes from—they buy from meat packers, not from the producers, and there is no way of tracing the original producer.)

Bill Niman is enthusiastic about Chipotle's marketing: "There has been a huge disconnect between farmers and end users, and these ads really explain what these 470 farmers are doing. And their customers responded very favorably. As soon as they raised the price and promoted the pork, their unit sales went up several-fold. That was pretty impressive. The restaurant trade took notice." Niman adds, "Every time they opened a new store, it enabled us to add another pig farm." That means many new Niman farmers, because there are now around 450 Chipotle restaurants, with more opening all the time. Ells says that some of his customers buy Niman for the taste and others because they want to support family farmers or deplore what factory pig farming does to animals. Whatever the motive, what matters for Ells is that they are supporting the overall cause of "Food with Integrity."

How far can a fast-food restaurant can go? None of Chipotle's meat can be labeled "organic" because the animals are not fed on organic feed or raised on land that has been certified as organic, so Chipotle calls its meat "naturally raised." The official U.S. definition of "natural" meat is so loose as to be almost meaningless—the animals can be confined indoors on bare concrete all their lives and fed chicken litter, antibiotics, and hormones, but the U.S. government says you can call their meat "natural" as long as you don't add anything artificial to the meat itself. Ann Daniels, Chipotle's "Director and Buyer of Gastronomic Delights," told us about Chipotle's stricter definition of the term: "We have defined 'naturally raised' to mean that the animals are never given any antibiotics or added hormones. Animal must also be humanely raised and treated. In the case of pork, the pigs are either on pasture or in deeply bedded pens. Gestation crates are not allowed." Chipotle visits suppliers to check that the animals are humanely treated. Ells makes some of the trips himself, visiting Paul Willis in Iowa to see Niman Ranch pigs two or three times a year. The chickens Chipotle buys get no antibiotics and are fed a purely vegetarian diet, with no slaughterhouse remnants. Ann told us that they have "more room per bird than in conventionally raised poultry houses." That, however, isn't saying much.

Chipotle buys Meyer Natural Angus beef, which is "Certified Humane," but the cattle are still kept in feedlots for 150 days and fed mostly on corn. Some of these feedlots—which, like the feedlots we visited and wrote about in Part I, are mostly near West Point, Nebraska—are very large, holding up to 20,000 cattle. The cattle get no antibiotics, slaughterhouse by-products, or added hormones, and about 10 percent of their diet is roughage, as compared to only 3 to 5 percent for in conventional feedlot beef.[1]

In summary, then, Chipotle uses pork from pigs raised on farms with good welfare standards, but there isn't a great deal of difference, from an animal welfare point of view, between the beef and chicken it serves and every other restaurant chain's standard factory farm products. Even these modest standards, Ells tells us, have yet to be implemented in all of Chipotle's restaurants. All of the pork Chipotle serves is naturally raised, but in July 2005, only about 100 restaurants were serving naturally raised chicken, and even fewer were serving naturally raised beef. By the time this book is published, naturally-raised chicken will have spread to Chicago, Minneapolis, and all of Florida. Ells says that the supply simply isn't available, as yet, to allow them to serve only naturally raised chicken across the entire country. "We think it's a better way to raise birds and results in better tasting chicken. We are committed to moving in that direction," he says.

You can always have your tacos and burritos vegetarian. About 15 percent of the black beans and pinto beans served at Chipotle are organically grown—up from 10 percent in 2003. Chipotle is still working out where it wants to go with organically-grown plant foods. They are interested in developing a protocol that will ensure their suppliers are practicing "sustainable agriculture," but may not correspond exactly to the current U.S. Department of Agriculture organic food standards. "We know we can't just flip a switch and have everything organic and free-range overnight," Ells says, "But we're committed to purchasing more and more sustainably raised food."

## A SUPERMARKET: WHOLE FOODS MARKET

John Mackey opened a store selling natural foods in Austin in 1978, when he was 25. He founded Whole Foods Market two years later, and in just five years the company had 600 employees. At that point, Mackey took some of them on a series of weekend retreats where they worked together on a "Declaration of Interdependence" that made all the employees stakeholders in the company and established the company's guiding principles—sell quality food, please customers, satisfy employees, create wealth, respect the environment, and conduct a responsible business.

Whole Foods ranks at 479 on *Fortune* magazine's list of America's 500 biggest businesses and has around 38,000 employees, 180 stores in the U.S., Canada, and Britain, and an annual turnover approaching $5 billion. Mackey is currently the company's Chief Executive Officer and Chairman of the Board. On one of his frequent visits to New York—where he had just opened Manhattan's largest supermarket—we met him at Gobo, a vegan restaurant in the West Village. He's a friendly, fit-looking man of around 50, who doesn't dress like the CEO of a major corporation.[2] Over dinner, we start with the things we have in common—philosophy, which Mackey studied at the University of Texas and still likes to talk about, and vegetarianism. Mackey told us he had been a vegetarian ever since he moved into a vegetarian co-op in Austin in his early 20s. He wasn't vegetarian at the time, and one of his motivations for moving in was to meet women he thought would be interesting. He did—and in doing so he drifted away from eating meat, but largely because he thought this was a trendy and healthy thing to do, rather than out of a strong ethical conviction. He continued being vegetarian for many years, primarily for health reasons and partly because of compassion for animals, but without a great deal of commitment.

Then in April 2003, Whole Food Market's annual meeting was picketed by Viva! USA, an animal rights organization that was calling for a boycott of Whole Foods. When Lauren Ornelas, Viva!'s director, got the microphone during question time, she blasted Whole Foods for the ducks it was buying from Grimaud Farm, in California. Although the farm had a reputation for producing high-quality ducks, fed on a diet free of animal by-products, antibiotics, and hormones, Viva! had discovered that the birds were reared in filthy, crowded sheds, had the tips of their bills cut off, and were denied access to water in which to immerse themselves—which ducks need to stay healthy. Mackey didn't react well.

"I actually said to Lauren, 'We have the best animal standards in the country—go bother somebody else.'" After the meeting, however, Ornelas came up to Mackey and they had a cordial exchange that led to an email correspondence. The line Lauren took, as Mackey later recalled, was, "Well, gee, Mr. Mackey, you're well-intentioned, but I don't think you're very well informed about the actual conditions of the animals." Mackey asked people at Whole Food Market who had visited Grimaud and was told that conditions there were "good," in comparison to their competitors, and since he was too busy to follow up every issue personally, that satisfied him, at least temporarily. After several weeks he sent Ornelas a final email that said, basically, "I'm done. We're not going to agree about this."

But Mackey was far from done. Over the summer of 2003, as he later told an interviewer, "I read a dozen books about how animals are raised in this country, going all the way back to Peter Singer's *Animal Liberation* in 1975. The more I read, the more I was interested in it. I said, 'Damn, these people are right. This is terrible.'" At that point, Mackey realized that he "couldn't continue to eat animal products—I just wanted to be a vegan." During his years as a vegetarian, Mackey said, "When it came to dairy products and eggs and that kind of thing, I just looked the other way . . . It's tempting to think: 'I'm doing enough . . .'" Now, after becoming a vegan, Mackey feels that "it's been a really good decision for me personally," and his values and his actions are aligned "one hundred percent."

Whole Foods, of course, is not vegan. Some might question how well that aligns with the CEO's values, but Mackey has a ready answer. "Whole Foods exists to meet the needs and desires of its customers, and not to pursue the personal philosophies of the founder/CEO, whatever those personal philosophies might be." If Whole Foods ignores the desires of its customers, he argues, it will go out of business. To support that claim, he cites research showing that only 10 percent of Whole Foods' customers are vegetarian and just 3 percent are vegan. Instead of committing business suicide by making the entire chain vegan, Mackey decided to educate Whole Foods' suppliers to produce their animal products in a more compassionate way, and to persuade its customers to make more compassionate choices.

After becoming a vegan, Mackey decided to visit Grimaud Farm himself. He found that the ducks had no access to outdoors, their bills were trimmed to prevent them from pecking each other, and, most stunning of all, they were not allowed to swim. He sent Ornelas an email telling her she was right—not just about ducks, but about chickens, pigs, and cows. He told her that Whole Foods would immediately begin using its influence and buying power to ensure that the meat it sells comes from animals who have been treated with a measure of dignity before being slaughtered. And he invited her to take part in the process. "I was at the office when that email came in," Ornelas says, "And I just about fell on the floor."

Ornelas wasn't the only one invited to help. Mackey contacted the Humane Society of the United States, People for the Ethical Treatment of Animals, the Animal Welfare Institute, and Animal Rights International, asking them to join him and other senior members of the Whole Foods team in meeting with his animal product suppliers and some leading animal scientists. From that meeting a working group began developing standards that will provide better lives for a wide range of animals. New standards for ducks, cattle, pigs, and sheep were posted on Whole Foods' Web site in November 2005, with others to follow. The next step is to certify that suppliers meet the new standards and designate their products with an "Animal Compassion" logo. That process should be complete by 2008. Then Whole Foods will inform its customers about the difference between the "Animal Compassion" standards and conventional ones. In that way, millions of Whole Foods customers will learn about the factory farms from which most animal products come—and about the alternatives to them. Already, Whole Foods does not stock fois gras, and since June 2004, it has not sold eggs from hens kept in cages. In 2005, it also eliminated such eggs from its kitchens and bakeries and from those it commissions to make baked or prepared foods.

Mackey takes part in the meetings that are drawing up the Animal Compassion standards, along with a dozen or more Whole Foods staff, including both of the corporation's co-presidents. The first species for which standards were drawn up was ducks, partly because that was the issue that Ornelas brought to Whole Foods, and partly because it seemed a good idea to start with a relatively small industry. Grimaud Farm had its people at the first meeting. After some discussion, they agreed that it would be possible to allow the ducks to go outside and even to provide a shallow pond for them to swim in—something that is unheard of in commercial duck production in America. To meet Whole Foods' standards, the ducks will also have to be able to search for food, so there must be some form of pasture or ground cover, not just bare dirt. There must be space for them to express natural behaviors including standing, spreading and flapping their wings, turning around, and preening themselves, all without touching another bird. Unlike many birds who end up as "poultry," the "animal compassionate" ducks cannot have their bills trimmed, their toes cut, or be mutilated in any other way. Breeding ducks must be able to establish nests that are comfortable and safe.

Mackey thinks it essential to create a viable alternative to factory farms. "Right now," he says, "Americans have to pretend factory farms don't exist. They turn their eyes away, because there's no alternative, there's no choice." In twenty years, he predicts, factory farming will be illegal in the United States. To make that day come faster, Whole Foods has set up an Animal Compassion Foundation, an independent, non-profit organization the mission of which is "to provide education and research services to assist and inspire ranchers and meat producers around the world to achieve a higher standard of animal welfare excellence while still maintaining economic viability." The foundation will provide a global exchange network for information on humane farming techniques.

In addition to the positive impact it is having on animal welfare, Whole Food Markets' huge sales of organic bananas led Dole to convert significant amounts of its Central American farmland to organic production, while the demand for organic eggs and dairy products helped the Organic Valley cooperative take on new organic farmers. At Whole Foods, all fruits and vegetables are clearly labeled "organic" or "conventional"—or occasionally, "transitional" to indicate a farmer who is using organic methods but has not yet achieved certification. Stores have some autonomy about what they buy and are encouraged to buy local food when it is available. Signs often give detailed information about where a product comes from and how it was produced—including, perhaps, details about how a chicken was raised and an invitation to visit the farm. Whole Food Markets also has a goal of making all its "own brand" products free of genetically modified organisms. To promote consumer choice, it labels its own products as free of GMOs when that can be verified and encourages other manufacturers to do the same. The chain was an early supporter of the Marine Stewardship Council's "Fish Forever" label, showing that fish came from a fishery that was operated in a sustainable manner. In 1999, Whole Foods discontinued the sale of Chilean sea bass (also known as Patagonian Toothfish) because the stocks were being overfished.

As that example shows, there are some choices that Whole Foods doesn't leave to its customers. The company takes its name seriously, and interprets it to mean foods that are not adulterated with synthetics.

Nothing sold at Whole Foods has artificial sweeteners, colors, or preservatives. If you want to buy a can of Coke, you are free to shop elsewhere. You can't buy your way onto Whole Foods shelves either, even if your product doesn't contain any forbidden ingredients. Unlike most big grocery chains, which take billions of dollars annually in "slotting fees" and "vendor allowances" from big food corporations wanting to get their products prominently placed, Whole Foods does not accept such payments.

Behind everything Mackey does at Whole Foods lies a larger philosophy about capitalism. In Mackey's view, a responsible business benefits not only its shareholders, but all its stakeholders, a term that includes customers, employees, suppliers, the local community, and the environment. (Animals, he told us, are stakeholders too, within the environmental category.) The community benefits not just from access to good food and more jobs, but because Whole Foods donates a minimum of 5 percent of its profits to non-profit organizations and gives employees time off with full pay—up to twenty hours a year—to do voluntary community service. The environment benefits through Whole Foods promotion of organic foods and also from in-house practices like putting solar panels on the roofs of its stores to power its lighting and reduce greenhouse gas emissions.

There is some debate about how well Whole Foods looks after its employees, because Mackey's libertarian outlook makes him philosophically opposed to unions. Instead of negotiating wages and benefits with unions, he gives his employees—or "team members" as they are called—opportunities to acquire shares in the company. All team members are eligible for stock options after three years' employment. In sharp contrast to the usual corporate model, 94 percent of the company's stock options go to non-executives. Mackey has a policy of "no secrets management" that enables any employee to see what any other employee is getting paid. No employee may be paid more than 14 times what the average employee gets, which is about $29,000, a good wage for the grocery industry. That policy limits Mackey's own salary to a little over $400,000 a year. Compare that with Wal-Mart CEO Lee Scott, whose 2003 salary and bonus, including stock grants, came to $17.4 million, or 966 times the salary of a full-time "associate," as Wal-Mart's store assistants are called.[3] At Whole Foods, teams of employees—for example, the seafood team, the prepared foods team, the customer service team—vote on whether to accept a new employee, after a trial period. They have an incentive to accept only good workers, because when teams perform well, everyone on the team gets a bonus. Mackey calls it "an organization based on love instead of fear," which sounds too good to be real, but whatever the system is based on, it seems to work: for eight consecutive years, Whole Foods has been listed by *Fortune* magazine among the best 100 companies to work for. In 2005, it ranked number 30.

The main objection to Whole Foods is the same point that Wayne Bradley made about Niman Ranch pork—it's too expensive. Almost every article about Whole Foods mentions its nickname: "Whole Paycheck." The *New York Times* headlined its story on Mackey "The $6 heirloom tomato." Mackey's response to criticism of his pricing is that Americans spend far less of their income on food than people in other countries do and "that's why most of it tastes so bad." He might also have mentioned that we spend a smaller proportion of our income on food now than we used to do—on average, only 6 percent of our total income goes towards buying groceries, down from 17 percent fifty years ago. In fact, we probably work for fewer hours to feed ourselves than people have anywhere, in all the millennia of human existence. In Mackey's view, if Americans want to eat better quality food, most of them have the means to pay for it.

In the organic movement, some say that by getting big, Mackey has "sold out" to the corporate world. He responds by asking why it is supposed to be bad to put an ethically responsible way of doing business into the mainstream. Whole Foods has shown that squeezing every last cent out of the production and distribution process isn't the only way to make good profits. Running an ethical business can do it too.

# PART III

# THE VEGANS

# 13

# JOANN AND JOE

JoAnn and Joe Farb are vegans: they eat no meat, poultry, fish, eggs, or dairy, and they avoid any food that contains derivatives of any animal product. This diet satisfies nearly all of their ethical concerns about farming and food. Many activists in the animal rights movement are vegan, rather than vegetarian, for they do not want to support any form of exploitation of animals. They don't wear wool or leather either—thereby showing that their stance is not merely based on concern for their own health. Vegans, however, often say that it is healthier, both for themselves and for the planet, to avoid eating animal products. Yet the vegan diet is controversial. That is in part due to the fact that whereas there have been vegetarians for thousands of years, veganism, at least on a large scale, is still relatively new. The Vegan Society in Britain, the world's first vegan society, was founded in 1944 and at that time had only 25 members. Its founder, Donald Watson, was a living advertisement for the diet he followed, dying more than sixty years later, at the age of 95. Now, in the United States, between 1 and 3 percent of the population—or somewhere between two and six million adults—report that they do not eat meat or fish, and about a third of that group say they are vegan.[1]

## VEGANS IN A BASTION OF CONSERVATISM

Named from the Shawnee word for "beautiful," Olathe, Kansas, lies on rolling hills and plains southwest of Kansas City, Missouri. Pioneers in covered wagons traveling westward on the Oregon and Santa Fe Trails often spent their first nights here. During the late 1890s, the town and the surrounding region supported radical farmer-labor populist groups, such as the People's Party and the Freedom Commune. Newspapers with titles like *Kansas Agitator* and Olathe's own *Progressive Thought and Dawn of Equity* thrived. Not today. In Thomas Frank's bestseller *What's the Matter with Kansas? How Conservatives Won the Heart of America,* Olathe is singled out as "a bastion of right-wing conservatism," even by the standards of solidly Republican Kansas, and a place that "consistently takes the most outrageous positions on the issues of the day."[2] Olathe is in Johnson County, home of most of the affluent suburbs of Kansas City. Eighty-nine percent of its 100,000 residents are white, four percent are African-American, and five percent are Latino. Only 4 percent of Olatheans live below the poverty line.

Joe and JoAnn Farb and their daughters, Sarina, 10, and Samantha, 6, live just west of Olathe, right across the road from the Prairie Center, a 300-acre preserve where, in summer, the prairie grasses that once covered much of North America still grow as tall as your living room ceiling. The Farbs are doing their part to conserve the original vegetation too—their fifteen-acre lot is a hilly savannah veined with walking paths through native grasses and groves of fruit trees and thorny Osage Orange. On the day we drive in, running late for lunch, two white-tail fawns are dawdling in the winding driveway, making us later still.

Joe and JoAnn walk out to greet us with a spring in their step. They look like a well-suited couple, less than average height, both trim, athletic, and, with full heads of dark hair—though hers is straight and his is curly. Joe is 52, a sales entrepreneur whose work takes him on trips around the Midwest. JoAnn, 41, is an author and a microbiologist, but she stays home to carry out the family's home-schooling plan for Sarina and Samantha. Sarina, who is "almost eleven," can often be found practicing the piano and working on her gymnastics routine. At the moment she is playing a piece by Haydn, but she's equally happy demonstrating a series of leaps and flips down a long, carpeted hallway. Samantha, who is only six, is learning to help in the kitchen and is, according to her mother, "a connoisseur of mud."

We sit down to lunch in a huge, high-ceiling kitchen with yards and yards of cabinets and counter space. In the center is a large island with a stove, an area for preparing meals, and a dining bar with high stools where lunch is set out. It will be the family's main meal today because the afternoon's activities will run late. Before we all sit down the family joins hands and sings:

> Back of the bread is the flour,
> and back of the flour is the mill,
> and back of the mill is the wind and the rain,
> and the Father's will.

"I made this salad," says Sarina, taking the tongs to drop a portion on her plate. "It's got Romaine lettuce, red cabbage, carrots . . . and onions and tomatoes."

"And I made the dressing," JoAnn says. "It's my basic: red wine vinegar, olive oil, some Dijon mustard, and herbs."

"And Samantha and I made this raw beet salad," says Joe, placing a couple of spoonfuls on JoAnn's plate. "Tell us what's in it, Sam."

"Beets, onions, and . . . garlic, and . . . ." She stops to giggle and take a drink of home-made lemonade.

"Olive oil and some parsley," JoAnn adds. "Sometimes we use dill."

Then a large bowl goes round the table and everyone takes two or three spoonfuls of a steaming rice dish.

"This is the family favorite," JoAnn says. "It's brown and wild rice stir-fried with tofu and red cabbage. I made it up once upon a time. I cut up some tofu into little cubes, then marinate it for a while in a mixture of soy sauce, lemon juice, garlic powder, marjoram, thyme, and cayenne pepper. Then I sauté the tofu and onions in olive oil with a bunch of fresh herbs and spices, throw in the cabbage until it's tender, throw in the rice—cooked first, of course—and it's ready to eat."

JoAnn became a vegetarian when she was 13. "My parents had a hobby farm down in the Ozarks and they put a few cows on it because they wanted beef that wasn't shot up with drugs and hormones. When they shipped off the cattle to be killed for food, that's when I first made the connection between meat and a living creature. It just didn't seem right to me. As soon as I made the connection, I said, 'I'm never eating a cow again.' I put mammals out of my diet immediately and considered myself a vegetarian."

It was a gradual transition from there. "In college I started cutting out other kinds of meat. But then I'd have turkey at Thanksgiving. Then somewhere into the science curriculum, I started to develop a whole different perspective on eating meat. I thought, 'Well, in the natural world there's this thing called survival of the fittest and it's OK to eat animals.' I didn't have an ethical problem with it anymore. But I was so used to not eating animals and I had so grossed myself out on the idea that meat was the guts or body parts of some living creature that the idea of eating meat still repulsed me. Avoiding meat was kind of a habit at that point, and health-wise I figured it was better."

When she was about 30, JoAnn worked for Merck, the giant pharmaceutical corporation. Her role, ironically, was to sell chemicals to the poultry and pig industries, and her work involved on-site visits. That led her to think again about the ethics of what she was eating: "I started to see how my personal food choices were the fuel that was driving the hideous things I was seeing: factory farms, the focus on chemicals and vaccines and antibiotics and all the drugs they're shooting animals up with. I saw that those things were only necessary because of the overcrowded, inhumane places the animals were kept in—to keep them alive under extreme living conditions. I read John Robbins' *Diet for a New America* and I went vegan pretty much overnight."

Joe says he grew up "a meat and potatoes man" but started cutting back on his meat eating for health reasons some time before he met JoAnn 12 years ago. "Then when I met her and started hearing about these animal factories she was going into, I became a vegetarian. Before we got married, we agreed to raise our kids vegan and to keep a vegan household."

Sarina comes into the conversation, saying that the vegan diet is "good for the animals." When asked about her other ethical concerns, she says, "Thousands of trees are being cut down to raise cattle. The rainforests are being cut down. And we're using lots and lots of water just to grow one pound of beef."

JoAnn explains that for her the issue is more the suffering than the death of animals. She talks of hunting, saying, "I find less of an ethical concern there than with the animals I saw in those factory farms." She adds that it may seem odd because she feels that she is not much of "an animal person." By that she means that she is not especially close to, or affectionate with, animals. She was when she was young, she says, but not since she has had her children. Joe adds that they have a companion animal—Maple, a cat who found his way to their doorstep a few years ago. The family has since adopted another homeless cat the same way. "If one finds its way to us," he says, "then we take care of it."

JoAnn still can't forget her visits to factory farms. "I remember walking through a swine facility one time where the workers had to wear respirators. OSHA required it because the foul air would damage their lungs. We had to walk through quickly and try to hold our breath. The odor is overwhelming, really sickening. And to think that those animals live in there their entire lives."

We talk about some of the other ethical concerns that hold the Farbs to a vegan diet. Environmental concerns are important, as Sarina said. So is a broad range of social concerns, such as labor issues, corporate responsibility, and food safety. Good parenting became a cause for her when she saw parents at vegetarian events doing things that concerned her—spanking, bottle-feeding, and "immersing their kids in an advertising-based culture and giving them crap to eat and not looking at the larger issues of health, nutrition, and lifestyle." So she wrote a book, *Compassionate Souls: Raising the Next Generation to Change the World*, to pull together all of those ideas. "It's about vegan parenting, of course, but more about the larger issue of how you create compassionate, good citizens with healthier values."[3]

"And of course what it ended up doing," Joe says, struggling to control his laughter, "was just pissing off everybody and their kids."

The conversation turns to Sarina's gymnastics—she's getting to a level where the best kids in the team are in the gym every day. Sarina can hold her own, even though she's the smallest member of the team, but she's not really serious about competing. Just the other day Sarina said to JoAnn: "Mom, I love gymnastics, but it's not my real life. My real life is activism." Sarina, it seems, feels a very strong desire to do something to make the world a better place. JoAnn and she have been reading together and then discussing biographies of scientists and humanitarians like Charles Darwin, Clarence Darrow, and Linus Pauling. It's easy to see that Sarina is unusually articulate and thoughtful for a child of her age.

JoAnn's reference to Sarina being the smallest member of the team reminds us to inquire about being a vegan mother and about how well a vegan diet worked for growing children. When JoAnn became pregnant with Sarina, her midwife made her concerned about her diet. "She really scared me. I started questioning if what I was doing was really OK, because she had been through the whole birth-pregnancy thing with a lot of women and I trusted her experience. So that sent me to the medical library, and I dug up everything I could find on vegan pregnancies and vegan kids. What I found reassured me—not only that what I was doing was safe, but that it actually had many advantages too. And my midwife did comment to me, after we went through the whole pregnancy and she had spent time at our house seeing how we ate, that the other vegetarians she had known did not eat the way we ate. Meaning, they were eating potato chips and soda pop as a substantial part of their diet."

When the children were born, JoAnn tells us, they were big babies—in the 90th percentile for weight. Obviously she was well-nourished during pregnancy. But later her children dropped right down to the 5th percentile, so they weighed less than 95 out of 100 children of their age. Looking at JoAnn and Joe, though, it's easy to see that this is more likely to be the result of genetics than diet. She is only 4 feet, 10 inches, herself, and he is 5 feet, 6 inches.

We asked how she makes sure the kids get enough protein. Back in the 1970s, Francis Moore Lappé's bestselling *Diet for a Small Planet* had spread the idea that to get enough protein from plant foods, you had to combine different kinds of food, since some, like beans and lentils, are low in one particular amino acid, while others, like bread or rice, contain high levels of that amino acid but are deficient in another one. Traditional ways of eating, like serving curried lentils over rice, or wrapping beans in a tortilla, or the Italian pasta and bean soup, turned out to be excellent ways of combining different forms of protein.

"You know when I was pregnant, I did do the protein combining, just to make sure I was covering all my bases, even though by then most experts were saying that was totally unnecessary. But I did do it. I tend to err on the side of caution. Now, I just make sure to include protein foods in every day's menu. Legumes, soy products. . . . We're pretty generous with nuts and seeds in our diet. I just try to keep the variety there, that's the main thing. And we're not getting empty calories. We just don't keep empty calories in our house. I think if you're not doing empty calories, then you don't have a lot of concerns." JoAnn does, however, make sure that the whole family gets enough vitamin $B_{12}$ by buying a vitamin supplement. In winter, because they don't get outside so much, they also take a multivitamin supplement that provides vitamin D.

We pick up on the remark about there being no "empty calories" in the house. Don't the kids ever crave candy or chips? "Sarina just got a goodie bag at the gym—the kids are always bringing goodie bags and giving them out there. So she brought home this goodie bag that had a bracelet in it that she wanted and all kinds of stuff that she didn't want: M&Ms, microwave popcorn, a Swiss Miss hot cocoa pack. I took it back up to the gym today, and just putting it in the car to take it up there, Sarina was like, "Oh that stuff smells bad! What is in it?"

"No soda pop, nothing like that?"

JoAnn gives a little laugh. "The closest they've ever come to soda pop has been at Vegetarian Summerfests every year, where they've had some supposedly healthier variety of soda pop."

"What about chips and stuff like that?"

"We do keep some baked organic corn chips on hand. And in a fix, if we're in a hurry or we're running out the door and we have to have food where we're going, I might bring some baked organic corn chips and some bean dip. That's what we do where some people would use McDonald's."

The family schedule does not permit more lunch conversation. It's time to do the grocery shopping. On the way, JoAnn stops briefly at Dillon's, a regional supermarket chain, to buy apples, oranges, and other fruit. She tells me that she is able to buy more and more organic produce now in this and other conventional supermarkets. The Farbs buy most of their food "in bulk" from Blooming Prairie, a co-op based in Iowa City, Iowa. They place an order about every six weeks and a truck brings the order to their house. But when supplies run low, they buy at Wild Oats—mainly for its organic produce.

"I buy organics for the health and well-being of my family, because I don't want to support farming practices that can pollute air, water, and soil, even far from where the food is grown, and because how I spend my money seems to be the most powerful voice I have in our corporate democracy." When we ask her to elaborate, she begins by saying that organic food means fewer chemical contaminants, and the way it is grown can also affect the nutrients in it. Then she adds that organic producers are not allowed to use genetically modified plants. These creations of modern biotechnology could, she believes, have "novel, potentially toxic, proteins"—but she's a microbiologist, and this starts to get quite technical, so we make a mental note to pursue it on another occasion when we can focus better on what she is saying. Instead we lead the conversation back to her other reasons for buying organic. "I'm vehemently opposed to corporations trying to patent or own life forms and then charge farmers royalties and make criminals of them just for saving their own seeds to regrow them the following year. For corporations it's just about making more money, but for some humans it's about survival. I really do try to vote with my dollar and not enrich those who are doing bad things in the world."

## Vegan Groceries

The Wild Oats store in nearby Overland Park is one of the newest in the national chain. Heaps of fruit and vegetables fill bins and tables, and there is plenty of floor space to get around the displays. Amazingly, our bulky tripod and video camera do not bring forth an angry store manager or any complaints from other shoppers. Later, when JoAnn goes through the register, the checker cheerfully mugs for our camera.

JoAnn begins picking out bunches of organic bananas. She gently rummages through a display of Romaine lettuce and picks out two heads that please her. Now she fills a plastic produce bag with organic red onions and holds it up, smiling, as if it's a trophy. Next, a bagful of organic kiwi fruit—it looks like about two dozen—and we move to a bank of products in a cooler along the wall. "Let me just show you some of the things I buy here," she says. She holds up a package of Soymage, a vegan soy-based cheese. "I'm looking at the label to see if it has isolated soy proteins, which are not organic." She picks up another brand and says, "I like this product better because it doesn't have any—it's called Veganrella and I buy the mozzarella and cheddar flavors."

We move on to the tofu section. She picks up a package. "I won't buy this product because it's not made with organic soybeans. There's so much soy grown in this country and much of it is probably genetically engineered. So it matters to me that the soybeans are grown organically." She puts it down and shows me another brand. "Now this is made with organic soybeans, but if you read the label carefully it says . . . I'm looking for the ingredients. It says . . . . Oh, they've changed it finally!" She looks up, pleased. "OK, this product is acceptable now. I had quit buying it completely because they were putting isolated soy protein in with the organic soy. Now they've quit doing that. Good."

She puts the package back and brings out two more packages. "I usually buy this brand. It's by White Wave and it's organic. I'm a little disturbed by the fact that they're a large corporate entity now, but I'll tell you why I like them: I had found this local guy making tofu, and I actually went there to see his process. He was a nice small business guy, but I watched him pour hot soy milk into Rubbermaid buckets and then add the coagulant. I didn't like the idea of hot liquid in plastic. So then I called White Wave and found out that they do all their processing in stainless steel. This is firm-style organic tofu. I always get the hardest tofu I can because they press most of the water out of it. I don't like to pay for water in my tofu."

She picks up another package from the display case. "I get this from time to time. This is a vegan parmesan cheese made with organic tofu and textured soy protein, which is not organic but we use this in pretty small amounts. Next, JoAnn gestures toward packages of "Smart Deli" vegetarian salami, turkey slices, and baloney. "Every once in a while I'll get some of these for my husband, who really likes this kind of stuff. But I never use it." She hands me a package of Smart Deli "roast turkey style meatless, fat-free slices." She reads the label to me. "Lightlife Foods. It has water, gluten, soy protein, natural flavors from vegetable sources."

Now we go to the freezer section. She opens a glass door and brings out a container. "This is the most awesome non-dairy ice cream I've found . . . yet, anyway. It's organic Soy Delicious . . . mint marble fudge. There are some other alternatives to dairy ice cream that are pretty good too. We also tasted the Rice Dream bars . . . but they're really high in fat. So we've kind of shifted away from their stuff. The Soy Dream desserts are awesome. I've taken these chocolate ice cream sandwiches to Sarina's friends in gymnastics class and they love them. They're called Li'l Dreamers. They're organic."

The Farb shopping cart is full. At the checkout, JoAnn runs back to get a big bag of organic lemons. Then she and the checker fill up JoAnn's cotton bags. We're off to park the groceries and watch Sarina finish up her gymnastics session. On the way we ask JoAnn if she and Joe have a lot of vegans in their social circles. "I wish we did," she says. "That is probably the worst part of being vegan right now. So much of our culture revolves around food—the raising of food, preparing it, and sharing it. We miss that community feel; we want to be able to share our food with lots of other people. But for us, food is a really big deal because of the foods that we will eat and won't eat. We do share our food with people, but other people can't really share their food with us. It's a challenge."

That leads us to ask her a more political question. We mention that we've read that this is a very conservative part of the country, always voting Republican, with a lot of fundamentalist Christians. "What's it like," we ask, "living here as a vegan with ethical concerns about the way we treat animals, when your neighbors are probably more outraged about abortion and gay marriage?" "It's quite lonely," JoAnn replies, "Oh, there are plenty of people here that I chat with daily—at the library, or the kid's activities, but that's not the same. It doesn't really feel like community to me. It's painful to be surrounded by so many people who just don't get any of what resonates so profoundly with every cell of my being. Many times I have told friends that I've always felt like an alien around here." So, after living seven years in Olathe and her entire life in Johnson County, she and Joe are thinking of moving to somewhere a little "crunchier." "Crunchier?" we ask, puzzled. She laughs and says that she means a community of granola-heads, but she's just kidding, what she really wants is a place with "more tolerance for differences, less emphasis on materialism, where people value creativity and are interested in working on issues related to peace and justice." (The family subsequently moved to Lawrence, the home of the University of Kansas.)

When we get back to the Farb home, JoAnn puts away her groceries and then pulls out bags of blueberries, bananas, and other fruit from the freezer. She puts it all in a giant blender with some soy milk and whips up enough smoothie to fill a half-gallon container. "I do this as often as I can to show Sarina's friends and their parents that healthy, organic food tastes as good or better than the junk treats they're used to." She giggles and makes a sly grin. "You want to bet they'll knock these down in no time?"

# 14

# GOING ORGANIC

Both the Masarech/Motavalli family and the Farbs prefer organically produced food. Jim and Mary Ann do some organic gardening, growing tomatoes, peppers, some berries, and herbs. They also buy organic meat, eggs, milk, and salad greens. JoAnn buys organic food, from lettuce to corn chips and soy ice cream, whenever she can. But buying organic food isn't, strictly speaking, part of being vegan. Some vegans have no preference for organic food, and some meat-eaters eat only organic food. But vegans are generally more aware of what they are putting into their mouths and more likely to buy organic food than the population as a whole.

In 2003, about 11 percent of Americans said they ate organic products daily, and 16 percent said they ate them weekly. Organic foods have now made the leap from "alternative" food co-ops and specialty health food stores to big supermarket chains—and not only chains like Wholefoods, Wild Oats, and Trader Joe's, but also Krogers, Super Target, King Soopers, and Price Chopper. Organic food is now available in about three in every four U.S. supermarkets. Yet overall, only about 2 percent of U.S. food sales are organic.[1] In the U.K. too, despite rapid recent growth, organic food still accounts for only about 2 percent of all grocery sales.[2] The mainstay of the organic movement is fresh fruit and vegetables, though sales of organic milk, eggs, cheese, and meat are growing rapidly. The range of processed organic food is also expanding and now includes pasta sauces, cereals, corn chips, ice cream, peanut butter, coffee, and even frozen dinners.

Worldwide, demand for certified organic products is increasing at 10 percent annually. But the amount of organic food produced is much larger than the amount that is certified, because many traditional farmers are already farming organically, although they may know nothing about certification schemes. Of farmers in developing countries who do learn of such schemes and apply for certification in order to export their produce under the organic label, about 80 percent do not need to change their farming methods.[3] The problem, from the perspective of the organic movement, is to persuade these traditional farmers to resist the pressure to adopt "modern" farming methods that will, organic advocates believe, prove costly in the long-run both for farmers and for the environment.

## WHAT IS ORGANIC FOOD?

Using the label "organic" to distinguish one tomato from another is a big stretch from the word's original meaning, for until the middle of the twentieth century it simply meant something living or derived from living matter. In that sense, the idea of an "inorganic tomato" is a contradiction in terms, unless it is, say, a tomato-shaped glass ornament. With very few exceptions—salt is one—all our food is "organic" no matter how it is produced.

The specific sense of "organic" we use when we speak of "organic food" today traces back to 1942, when J. I. Rodale launched a magazine called *Organic Gardening*. Nowadays Rodale is hailed as a pioneer, but then he was often derided as a crank and a throwback to obsolete ways of farming. He advocated maintaining soil fertility and stability by putting organic matter—animal manure or compost—back into the soil rather than relying on the "inorganic," or synthetic, fertilizers that were then widely seen as the modern way to go. So in Rodale's usage, it was the fertilizers, and from them, the farming methods, rather than the food, that were organic, and the concern was primarily with the soil, not with issues like biodiversity or animal welfare. But the meaning of "organic farming" soon parted company from Rodale's original narrow distinction between fertilizers. Varying definitions spun out of control as different associations of "organic farmers" tried to set standards in accordance with their own values. Some wanted to stick with a narrow definition in terms of what you could and could not put on the soil, the crops, or the animals. Others wanted to include an entire way of life, including healthy living, an equitable form of distribution, concern for wildlife, and so on. Among organizations of organic farmers around the world, the broader view prevailed. The International Federation of Organic Agriculture Movements settled on this definition:

> Organic agriculture is an agricultural system that promotes environmentally, socially, and economically sound production of food, fiber, timber, etc. In this system, soil fertility is seen as the key to successful production. Working with the natural properties of plants, animals, and the landscape, organic farmers aim to optimize quality in all aspects of agriculture and the environment.[4]

Such a definition does not, however, lend itself to being reduced to a label that can be put on products to show that they were produced organically. Without specific standards that could be encapsulated in a label, consumers were often unsure what the various "organic" labels used by different associations and producers really meant.

In the United States, food labelled "organic" is now legally required to meet a standard set by the U.S. Department of Agriculture. In Europe the term is defined by European law and producers using the word "organic" to describe their products must hold a licence from an approved certifying body. These bodies – like the Soil Association, the Organic Food Federation, or Demeter – do inspections and spot checks to ensure that the producer is meeting the required standard. The standards require farming methods that are sustainable in the long term. No synthetic fertilizers, herbicides or pesticides are permitted, although some naturally occurring pesticides may be used. No genetically modified plants or animals can be certified as organic. Animals must be fed only organically grown feed, without antibiotics, steroids or hormones, and must be able to go outside, generally on pasture, during daylight hours. The certifying bodies may also set their own standards above the minimum baseline required by E.U. law. The Soil Association, for example, has higher standards in many areas, including animal welfare.

## WHY BUY ORGANIC?

### 1. Health

People buying organic food want to avoid unnecessary risks, and they believe that more natural methods of producing food are likely to be healthier. That belief received powerful reinforcement from the outbreak of mad cow disease in Europe and the consequent revelation that intensively farmed cattle are fed slaughterhouse remnants. From eating the meat of these cattle, at least 150 people—and some say many more—contracted a slow-acting, fatal disease. No wonder millions of consumers decided that the old ways might be safer, especially where their children are concerned. In Britain, organic baby food now accounts for half of all the baby food sold, while the German baby food market has been described as "on its way to becoming more or less exclusively organic."[5]

Organic food contains fewer pesticides. Drawing on data covering 94,000 food samples, a Consumers Union research team found 73 percent of conventionally grown foods and 90 percent of conventionally grown apples, peaches, pears, strawberries, and celery had pesticide residues, as compared with only 23 percent of organically grown samples. Where the same pesticide was found in both conventional and organic foods, the levels of the pesticide were significantly lower in the organic food.[6] Scientists at the University of Washington tested the urine of children eating a conventional diet and children eating predominantly organic produce and found that these differences in pesticide levels are detectable in our bodies. Some of the children on a conventional diet had pesticide byproducts in their urine that indicated an intake of pesticides above the "negligible risk" level recommended by the Environmental Protection Agency's guidelines. The children who ate organic foods had a median level of pesticide byproducts only one-sixth that of children eating conventionally farmed foods, suggesting that their intake of pesticides was well within EPA recommended limits.[7]

Yet the British, French, and Swedish government food agencies have all recently concluded that there is no scientific evidence that organic food is safer or more nutritious than conventionally produced food.[8] Michael Pollan, who has written for *The New York Times* on farming, writes: "The science might still be sketchy, but common sense tells me organic is better food—better, anyway, than the kind grown with organophosphates, with antibiotics and growth hormones, with cadmium and lead and arsenic (the E.P.A. permits the use of toxic waste in fertilizers), with sewage sludge and animal feed made from ground-up bits of other animals as well as their own manure."[9]

## 2. The Environment

For the Farbs, and for Jim Motavalli and Mary Ann Masarech, supporting a method of farming that does not seriously pollute the air, water, and soil is an important reason for buying organic. Jim and Mary Ann also point to the waste of energy and water that is involved in conventional farming. But to what extent does the label "organic" indicate that less damage was done to the environment in producing that item than was done in producing similar items by conventional methods?

Much depends on the type of farms we are comparing. As we have seen, large factory farms raising chickens, pigs, or cattle can cause severe air and water pollution. But organic farms also can, as we shall see later in this chapter, concentrate animals together and pollute significantly. The organically raised animals will not pass steroids through their bodies and into streams, because they do not receive steroids. Five thousand cows fed on organic grain, however, will create the same amount of manure as that number of cows fed on conventionally produced grain. Granted, an organic egg producer can't get as many hens into a shed if they are all on the floor rather than stacked up in cages. But if organic farming continues to grow, there is nothing in the present regulations to stop a progression to even larger organic farms with even more animals. So the label "organic," at least as used in the USDA regulations, is no guarantee that a product comes from a farm in harmony with its environment.

Nevertheless, as long as we avoid sweeping generalizations that imply organic farmers can do no wrong, there is no real dispute that, overall, organic farming is better for the environment than conventional farming. A workshop held under the auspices of the Organization for Economic Co-Operation and Development, hosted by the U.S. Department of Agriculture in Washington, DC, in 2002, concluded, after four days of discussions involving 140 experts from 22 different countries, that "the strong balance of evidence from research, field trials, and farm experience is that organic agricultural practices are generally more environmentally friendly than conventional agriculture, particularly with regard to lower pesticide residues, a richer biodiversity, and greater resilience to drought. Organic farming systems also hold the potential to lower nutrient run-off and reduce greenhouse gas emissions."[10]

Here are some of the more important environmental benefits of organic farming.

**I. Organic farming maintains the quality of the soil.**

When a virgin field is tilled and then fertilized with synthetic fertilizers, it will lose between 50 and 65 percent of its nitrogen and soil carbon over fifty years.[11] After that, increasing inputs of fertilizer—and thus of fossil fuel energy—will be needed to maintain yields. If that no longer pays, the land will be abandoned, becoming a wasteland on which little grows.

Organic farming has a different philosophy. It sees farmers as stewards of the land, harvesting its fruits while they care for it so that they can leave it to future generations in a condition as good as, or better than, it was when they started farming. So organic farmers maintain and enrich the soil by adding organic matter. That increases the number of worms and micro-organisms.[12] Soil rich in organic matter needs less irrigation because the soil holds moisture better. It is also less likely to blow away in the wind, or wash off with every storm. A study of two adjacent wheat farms on similar soil near Spokane, Washington, found that over a 37-year period, the conventional farm lost more than 8 inches of topsoil, while the organic farm lost only 2 inches. The scientists concluded that the productivity of the organic farm was being maintained, while that of the conventional farm was being reduced because of high rates of soil erosion.[13]

**II. Organic farming fosters biodiversity.**

The expansion of intensive modern agriculture, with its monoculture crops and intense use of pesticides and herbicides, threatens endangered species. Rare plants are indiscriminately sprayed with herbicides, along with more common weeds. Insecticides eliminate the prey of many birds, and small mammals may be poisoned too. Organic farms, in contrast, use no herbicides, fewer pesticides, have more organic matter in the soil, and tolerate hedges or other uncultivated areas. All this makes them a haven for endangered species of plants, insects, birds, and animals. In a survey of the evidence published in the journal *Biological Conservation* in 2005, scientists reviewed 76 separate studies comparing the impact of organic and conventional farms on such things as plants, soil microbes, earthworms, spiders, butterflies, beetles, birds, and mammals. They found that the majority of these studies demonstrated that the abundance and richness of species tends to be higher on organic farms. Significantly, the differences applied particularly to species that have experienced a decline because of the intensification of modern agriculture.[14] In 2005, a five-year, government-funded study of British organic farms gave further support to that conclusion. [15]

**III. Organic farming reduces pollution from nitrogen run-off.**

Conventional agriculture relies heavily on synthetic fertilizers, especially nitrogen. World-wide, the use of nitrogen as a fertilizer has increased tenfold in the last fifty years. Half to two-thirds of this nitrogen makes its way into rivers and other ecosystems, affecting both freshwater and marine environments.[16] The most dramatic result is the dead zone in the Gulf of Mexico. Like the dead zone in Chesapeake Bay described in Chapter 2, the Gulf of Mexico dead zone is caused by too much nitrogen, but here the dominant source—56 percent, according to the U.S. Geological Survey—is chemical fertilizer run-off rather than animal manure, which contributes 25 percent. The Gulf of Mexico dead zone has grown dramatically over the past twenty years, and when it peaks each summer, it now covers an area larger than the state of New Jersey. The peak comes a month after the spring use of nitrogen fertilizers in the Midwest Corn Belt—a month is the time it takes for the water from the Upper Mississippi to reach the Gulf. The expanding dead zone is disrupting fishing. This is only one of 146 dead zones around the world, and not even the largest—that is in the Baltic Sea. Nitrogen fertilizer runoff is largely responsible for most of these. Forty-three of the dead zones occur in U.S. coastal waters.[17] A shift to organic farming, which does not use synthetic fertilizers, would dramatically reduce water pollution from nitrogen, and so shrink the dead zones.

**IV. Organic farming avoids the heavy pesticide and herbicide use typical of conventional farming.**

Conventional farming relies heavily on pesticides, including insecticides and herbicides. Pesticide use per acre more than doubled between 1931 and 1997, although it has decreased slightly since then.[18] During the 1990s, the U.S. Geological Survey collected more than 8,000 water and fish samples across the country and analyzed them for 76 different pesticides. Some key findings were:

- More than 90 percent of water and fish samples from all streams contained one, or more often several, pesticides.
- About half of all groundwater samples contained one or more pesticides.
- The highest rates of detection were for the most heavily used herbicides—atrazine, metolachlor, alachlor, and cyanazine—which were common in streams and shallow ground water in agricultural areas.
- Levels of any individual pesticide exceeding drinking water guidelines set by the Environmental Protecton Agency were found in only 1 percent of samples, but there is uncertainty about the risks of low-level exposure to multiple pesticides. Moreover, for about half the pesticides detected the EPA has not set any guidelines.
- Close to half of the agricultural streams sampled had pesticide levels that exceeded Canadian guidelines for the protection of aquatic life. (The report referred to Canadian guidelines because there are no U.S. guidelines for this purpose.)
- Pesticides like DDT and Dieldrin, which have not been used since the 1960s, were still present across the country. DDT was found in almost every fish sample.[19]

Organic farmers are permitted to use only a very limited range of insecticides, selected because they are natural products or their safety is well-established. Hence, organic farms will not, to the same extent as conventional farms, release insecticides into the air or nearby rivers. They are not permitted to use any herbicides at all.

**V. Organic farming uses less energy for a given yield than conventional farming.**

Organic farms do not use synthetic fertilizers, the manufacture of which requires a lot of energy. According to a study funded by the British Department for Environment, Food and Rural Affairs, organic crops used 35 percent less energy per unit of production and organic dairying 74 percent less.[20] Scientists at the University of Essex found that organic farmers in a range of different countries required only 30 to 50 percent of the energy consumed in conventional farming systems.[21]

### VI. Organic farming stores more carbon in the soil, thus off-setting carbon dioxide emissions.

Organic farming increases the amount of organic matter in the soil—matter that would otherwise rot above ground and produce carbon that would go into the atmosphere. So if organic farming spreads, that might reduce the severity of climate change. But how great an advantage organic farming has over conventional farming here is controversial. The Rodale Institute has carried out a 23-year study of the amount of carbon stored in the soil of its model farm and calculated that if the organic methods it uses were applied on all the cropland in the United States, 580 billion pounds of excess carbon dioxide could be sequestered in the soil every year. That's about four times the quantity of emissions that would be saved if the fuel efficiency of all cars and light trucks on U.S. roads were doubled.[22] But questions can be raised about how long annual carbon savings could continue, since eventually the organic matter will decompose and release carbon back into the atmosphere.

In any case, much depends on what forms of organic and conventional farming are compared. Organic farmers who use a lot of compost and animal manure and periodically grow cover crops and plow them into the soil will have much more carbon in the soil than farmers who use only synthetic fertilizers and till their soil, because tilling leads to a reduction in organic matter. But the rules for being certified "organic" in the U.S. do not require compost, manure, or cover crops. They simply make the use of such techniques more likely by banning most other methods of keeping the soil fertile. The ideal organic farmer who practices the methods used by the Rodale Institute is making a significant contribution to reducing the carbon build-up in the atmosphere, and the standard conventional farmer is not. Buying food labeled "organic," however, does not guarantee that it was grown in keeping with the Rodale Institute methods.

There are two offsetting factors relative to climate change and organic farming to consider. It is often claimed that conventional farming produces higher yields per acre, on average, than organic farming.

Therefore, if we need to produce a given quantity of food, we might use less land to produce it if we use conventional methods. Suppose we then took this extra land and planted it with trees, as part of an agro-forestry project. According to some estimates, trees absorb about eight times as much carbon per acre as soil can, even organically cultivated soil. That suggests an alternative strategy for storing carbon: grow the food we need by conventional methods on fewer acres, and plant trees on the rest.[23] Of course, this presupposes that conventional farming really does produce higher yields than organic methods. The Rodale Institute conducted a 22-year comparative trial of conventional farming and organic farming. Although the yields from conventional farming were higher in the short-term, over the entire period of the trial, corn and soybean yields were just as high on fields farmed organically.[24]

The second factor relative to global warming essentially concerns cows burping and farting. Cows produce methane, a greenhouse gas that is at least 20 times more potent than carbon dioxide and, overall, is responsible for about 2.5 percent of the total effect of greenhouse gas emissions. Cattle may be responsible for close to half of the world's methane emissions. They produce more of it when they eat foods with a high fiber content, like grass or hay. Organic cows generally eat more grass and hay than non-organic cows. Moreover, organic cows do not get bovine growth hormone, or BGH, which increases a cow's milk yields by about 10 percent. That means that you need 10 percent more organic cows to produce a given quantity of milk, and the extra cows will put more methane into the air.[25]

## TINKERING WITH GENES

For JoAnn Farb, one important reason for buying organically produced food is that organic producers are not allowed to use genetically modified plants, also known as GM plants, or GMOs, for genetically modified organisms. GMOs have had their genes altered using recently discovered techniques. Sometimes an existing gene may be removed, or its operation blocked. In other cases, a gene is inserted from another organism, sometimes from an entirely different species, to confer a desired trait on the plant or animal. If a different species is used, the result is known as a transgenic organism. Through transgenic manipulation, scientists can introduce traits that could not be bred into the plant or animal by traditional breeding methods. The debate about GMOs has become extraordinarily polarized, with proponents hailing them as the solution to the global food problem, and opponents labeling them "frankenfoods," to bring to mind Mary Shelley's novel in which Dr Frankenstein creates a monster that gets out of control.[26]

To date, despite a lot of talk in the 1990s about how genetic engineering could improve the productivity of farm animals, no genetically modified animal has been approved for commercial food production. As a 2002 National Research Council Report noted, the success rate for producing transgenic pigs, cattle, sheep, and goats ranges from zero to 4 percent. Most of these animals die well before maturity, and even among the survivors, "many do not express the inserted gene properly, often resulting in anatomical, physiologic, or behavioral abnormalities."[27] Some of this research has caused considerable suffering in these abnormal, sick, and short-lived animals. Nevertheless, research is continuing, and news media frequently carry stories saying that some form of genetically modified animal will soon go into commercial production.

One prospect is faster growing fish who will make aquaculture more efficient. The big problem here is the risk of the GM fish escaping and interbreeding with native fish, with unpredictable consequences for marine ecology. As we saw in Chapter 9 when fish farms are located in the sea, escapes are frequent and can involve very large numbers of fish, so the risk is substantial. Land-based aquaculture would offer greater security, but for many species is prohibitively expensive. Attempts to make the growth-enhanced fish sterile have so far not met with the 100 percent success that would be required to eliminate the risk of interbreeding. Hence the immediate future of GM fish is uncertain.[28]

Another GM animal that has caused considerable interest is the so-called "Enviropig," developed by researchers led by Professor Cecil Forsberg at the University of Guelph, in Canada, to overcome two problems with the intensive rearing of pigs. Pigs fed on a diet of corn or wheat are unable to absorb a form of phosphorus that these cereals contain. Farmers therefore add phosphorus to their diet. This adds to the cost of pig production and makes the manure the pigs produce very high in phosphorus. If this manure is repeatedly applied to farmland, the phosphorus will eventually run off and pollute streams and groundwater, causing excessive growth of algae and killing fish. The "Enviropig" has been given a gene that enables it to absorb more phosphorus in its stomach, thus eliminating the need for supplemental phosphorus, while at the same time reducing the level of phosphorus in the manure by up to 60 percent. The meat of these pigs, however, has yet to be approved for human consumption.[29]

Genetically modified plants, on the other hand, have achieved rapid commercial success. In 1996, there were only four million acres planted with GM crops, worldwide. Seven years later that had risen to 167 million acres. The most widely used modifications, accounting for 99 percent of all GM crops sown, confer resistance to pests, or make the plant tolerant to herbicides, enabling farmers to spray their entire fields, crops and all, with a herbicide that will kill weeds but not the crops.[30] In the U.S. in 2004, 85 percent of all soy sown was genetically modified, as were more than half of all canola, about half of all papayas, and 45 percent of all corn. In Australia, 30 percent of all cotton grown is genetically modified, and there is some genetically modified canola. GM corn, soy and potato products may be imported into Australia. The United States is by far the world's largest producer of GM food, accounting for nearly two-thirds of global plantings of GM crops. Adding in Brazil, Argentina, Canada, China, and South Africa takes the figure to 99 percent.[31] In contrast, despite claims that GM crops are needed to feed the world's growing population, only about 2 percent of arable land in developing countries is sown with GM crops.[32]

In the European Union, Russia, China, Japan, Korea, Thailand, Australia, and New Zealand, products containing GM foods must state that on the label. In the United States, there is no such requirement, despite a 2003 ABC news survey showing 92 percent support for mandatory labeling of GM goods. American consumers who do not buy organic foods are almost certainly eating some GM foods, although they have no way of knowing which of the foods they buy contain modified ingredients.[33] According to Stephanie Childs, of the Grocery Manufacturers of America, there are GM ingredients in roughly 75 percent of U.S. processed foods, like breakfast cereals, other products made from grains, frozen dinners, and cooking oils.[34] JoAnn does her best to avoid GM foods by rejecting anything that contains corn or soy, unless it is organically produced.

When JoAnn first mentioned her opposition to GM foods, she spoke about "novel, potentially toxic, proteins" that could be created by genetic modification. When we followed up on this, she said that maybe the proteins she had in mind were not entirely novel, because they did occur in nature, but only in trace amounts. As an example, she mentioned one of the most widely grown GM plants, "Bt corn," a form of corn that has been given a gene from a bacterium, *Bacillus thuringiensis*. This bacterium has its own inbuilt pesticide and so Bt corn does not need spraying against corn borer, a common pest that does significant damage to normal corn. This modification makes the corn plant produce, in every cell, a protein from the bacterium that is toxic to moths. When we eat Bt corn, we also eat the protein, and, as JoAnn puts it, "There isn't any way in the natural world that we would have encountered that quantity of this substance." Then she asks, rhetorically: "Why would I want to put into my body potential toxins that our species has had no biological experience with?"

Lee Silver was a professor of molecular biology at Princeton University before he became interested in the public policy questions surrounding the application of advances in the biological sciences. That interest led him to switch disciplines, and he now has the title of Professor of Molecular Biology and Public Affairs. His recent book *Challenging Nature* defends genetic modification of plants and animals.[35] We asked him if JoAnn was right to say that when we eat Bt corn, we are eating larger quantities of a protein toxic to moths than we would if we avoid genetically modified foods. "Yes and no," he replied, and went on to explain that it is true that Bt corn does contain the protein, called cry1Ab, and non-GM corn that has been grown with synthetic pesticides—or without any pesticides—does not. Ironically, however, it is permissible for organic farmers to spray their corn with *Bacillus thuringiensis* bacteria in response to insect infestation. So the cry1Ab protein could be present in organically-grown corn.

Professor Silver then added three other reasons why he was not worried about any health risk from Bt corn. First, although Bt corn is approved for human consumption, most of it is fed to animals, or made into corn syrup and used as a sweetener in soft drinks and other food. Since it contains no nutrients other than carbohydrates, corn syrup does not contain cry1Ab, or any other proteins. Second, according to the data Monsanto provided to regulatory agencies, the amount of cry1Ab in the GM corn itself is 3 parts in 10 million, which Silver described as "more than zero but pretty minuscule." Third, the cry1Ab protein evolved in bacteria to bind to a very specific receptor in the gut of certain insects.

Vertebrate animals don't have this receptor, so the protein can't have any effect on vertebrates—or at least, Silver says, "that's the theory, and it's supported by tests performed in the lab."

Most scientists share Silver's view. "Scientists generally agree," states a report from the United Nations' Food and Agriculture Organization, "that the transgenic crops currently being grown and the foods derived from them are safe to eat"—although the report then disconcertingly admits that "little is known about their long-term effects." An expert committee reviewed about fifty studies of GM food for the International Council for Science and found that after many millions of meals containing GM food had been eaten, "no demonstrated adverse effects" were found. But it too had a caveat, conceding that its finding about currently available GM foods "does not guarantee that no risks will be encountered as more foods are developed with novel characteristics."[36] Silver also emphasizes that the safety of one particular GM food, like Bt corn, does not tell us anything about the safety of other GM foods: "Each GM modification has to be considered on a case-by-case regulatory basis," he says. "I would want to see the theory and empirical data on any other GM crop before making any claims about safety."

Ethical arguments against GM crops that go beyond issues of personal health are less often heard in the U.S. than in Europe, where there has been far greater public concern about GM food. The most fundamental ethical objection is that genetic modification is a form of human arrogance, almost like playing God. It is wrong, some say, for one species to tamper with the nature of other species by, for example, inserting a gene from a fish into a plant, so as to create an entirely new kind of plant. The second major argument is that GM crops pose an unacceptable risk of irreversible environmental damage.

The first of these arguments can take either a religious or a non-religious form. If life and all the species are seen as God-given, then for humans to set about to modify them may be regarded as a blasphemous attempt to improve upon God's creation. Without invoking religious beliefs, a similar argument can be put in terms of the intrinsic value of nature and the belief that we should not alter it. Nature itself, rather than God, then becomes sacrosanct. But it isn't easy to see why both the religious and the non-religious forms of the argument should not also rule out the kind of selective breeding that has, over many thousands of generations, transformed wild animals into the familiar domestic animals we have today. Was it blasphemous for humans to transform Burmese jungle fowl into the modern chicken? If GM corn is "unnatural," so too is a turkey with a breast so large that it can only reproduce through artificial insemination. Why should one way of changing species be "playing God" or contrary to nature and the other not? Unless we are to turn our back on the domestication of plants and animals and revert to being hunter-gatherers, we cannot logically hold that interfering with the nature of species is intrinsically wrong.

Nevertheless, we still have a lot to learn about natural processes, whether it is a question of genetics, our own health, or our planet's ecology. Awareness of what we do not know might properly make us cautious about creating novel organisms. One of the risks is to the environment. Tallying the environmental scorecard for GM plants, however, is not straightforward. GM plants do offer some ecological benefits, like reduced use of synthetic pesticides with Bt corn and Bt cotton. Against this, it has been claimed that the biological pesticides threaten harmless, and sometimes endangered, insects. But do they? The story of the monarch butterfly and Bt corn is, in the view of GM advocates, a case study in public misinformation.

The monarch butterfly is a particular favorite of conservationists, because of their extraordinary pattern of migration. Each year, in late summer, millions of butterflies begin traveling from the northern U.S. and Canada to a few specific locations in Mexico and California, where they stay over winter. The buttefly populations become so dense in these places that they are a tourist attraction. The ability of the butterflies to find these sites is baffling, because they have never been there before—they hatched from eggs laid thousands of miles to the north, several generations on from eggs laid by butterflies who began the return migration but died along the way, leaving it to their offspring to complete the journey north.

In recent years, however, the number of butterflies arriving in Mexico has dwindled. In 1999 John Losey, an entomologist at Cornell University, published a paper in *Nature* showing that pollen from Bt corn kills the caterpillars from which monarch butterflies develop. That finding was seized on by opponents of GM crops. But it was based on laboratory tests in which the Bt pollen was placed on milkweed, the favored food of the caterpillars. Though caterpillars coming into contact with Bt pollen in this way will die, subsequent field studies conducted by six teams of scientists and published in 2001 in *Proceedings of the National Academy of Sciences* showed that corn and milkweed are not generally found in the same field, and in any case the amount of pollen consumed by the caterpillars was not likely to be toxic. The risk of harm to the monarch butterfly from Bt corn was therefore, the studies concluded, very small in comparison to the risk from conventional pesticides, or droughts. As so often happens, the reassuring field studies received far less publicity than the original, more alarmist, finding.[37]

The use of herbicide-tolerant GM crops has also been questioned, for it allows herbicide to be sprayed less discriminatingly, which could have a deleterious effect on wild plants and invertebrates. On the other hand, using herbicides, rather than tilling the soil, has benefits for soil conservation and the ability of the soil to retain carbon. So the picture regarding both pesticides and herbicides is a mixed one and does not obviously come down against GM crops.

The most serious concern, however, is that GM plants will cross with wild relatives. Just as the introduction of new species into an ecosystem can create an ecological tragedy—the introduction of a few pairs of rabbits into Australia soon produced a population that eroded grasslands and drove out native species—so the introduction of a new GM plant could trigger a new ecological disaster, impossible to control. If insects have been keeping a vigorous wild plant in check, and that plant now gains the Bt gene from a GM plant and becomes resistant to insects, it could turn it into a rampant new weed that chokes other plants, possibly with devastating ecological consequences.

This is not a far-fetched scenario. It has been shown that pollen from GM plants does spread beyond the borders of the fields in which the GM plants are growing, and that it can fertilize wild relatives of those plants, thus causing what scientists call "gene flow"—passing the modified genes of the GM plants on to wild plants. To date, there is no evidence that this has caused any wild plant to become more invasive. But our luck may not always hold. The Food and Agriculture Organization's report on biotechnology acknowledges that the complete isolation of GM crops is "not currently practical," but suggests that gene flow can be minimized by management techniques such as using buffer zones and avoiding the planting of GM crops in areas where their wild relatives are present.[38] That seems naïve. If GM crops produce greater profits for farmers than conventional crops, and regulations are absent or not strictly enforced—a familiar story in some countries where GM crops are now being grown—then some farmers will sow the GM crops wherever they can. It only takes one instance of gene flow conferring a significant advantage on an invasive plant and its offspring could spread, like Australia's rabbits, across an entire continent.

For JoAnn Farb, distrust of the U.S. regulatory process is an important reason for avoiding GM foods. "Barely a week goes by," she told us, "that I don't see some article in the newspaper showing how the FDA (Food and Drug Administration) continue to approve things that later turn out to be very troublesome. And then we find out that a majority of the people that are regulating these things have financial ties to the very companies that are submitting the things to be approved. I don't trust the process at all." We asked Lee Silver what he thought about JoAnn's distrust of the regulatory process. He said that he found that a difficult question to answer, in part because it was not one on which he had any scientific expertise. He acknowledged, though, that there is "ample evidence" to warrant skepticism about the regulators.

There is also the separate question of whether, even if the regulators make the right decisions, the corporations will comply with them. In 2000, it was revealed that Starlink, a Bt corn approved only for animal feed because of concerns that some humans may be allergic to it, was found to have been harvested and made into corn flour for human consumption. The flour was later found to have reached Mexico, Japan, and Europe, as well as the U.S.[39] In 2005, Syngenta, one of the world's largest agricultural biotech companies, revealed that for four years it had been distributing a strain of genetically modified corn that did not have regulatory approval. Syngenta was later fined $375,000 for the regulatory breach. The corn was sown over about 150 square kilometers of farmland. Initially, Syngenta said that the corn, known as Bt10, was safe and virtually identical to another variety, Bt11, which had been approved. A few days later it was revealed that the Bt10 variety also contained a marker gene for resistance to ampicillin, a widely-used antibiotic. Such marker genes are used to identify certain strains, and are usually removed before release. This one was not.

A Syngenta spokesperson said that the gene was not mentioned initially because it "wasn't relevant to the health and safety discussion," but at least one expert body, the European Food Safety Authority, which advises European Union governments on food issues, does not consider the inclusion of such a gene irrelevant. Just a year earlier, it had said that marker genes conferring resistance to ampicillin "should be restricted to field trials and not be present in genetically modified plants placed on the market." Michael Rodemeyer, director of the Washington-based Pew Initiative on Food and Biotechnology, said that the release of the unapproved corn was an indication of the lack of any thorough monitoring system for genetically modified foods in the U.S.

There is, therefore, a case for avoiding genetically modied foods, because we should not support growing crops or releasing animals where there is even a very slight risk that this could cause an environmental disaster. To take this view is not to say that it is, in principle, wrong to modify any organism, or even to say that we will not one day have safe GMOs that bring sufficient benefits to justify commercial use. An effective regulatory system should consider each case on its merits. Since the developed countries can already produce an abundance of food without using GM techniques, we do not need to take big risks to produce more food.

## WILL GM HELP THE POOR?

What about the developing world, where a billion people cannot be sure they and their families will have enough to eat? Would it not be justified to take a greater risk for their sake? By transferring genes to existing food plants we will, it has been claimed, be able to produce plants with increased yields, greater drought resistance, and the ability to tolerate soils with a high salt content. Crops that produce proteins toxic to insects could reduce pesticide poisoning among farm workers, as well as pollution from synthetic pesticides, while plants with reduced needs for water and fertilizer could make farming more environmentally friendly. In addition, genetic engineering could even make staple foods more nourishing. More than 200 million people suffer from vitamin A deficiency, which is responsible for an estimated 2.8 million cases of blindness in children under five years of age. For many of these people, rice is their staple food.[40] German and Swiss university researchers therefore teamed up to genetically engineer a rice—known as "golden rice" because of its distinctive color—that produces beta-carotene, the precursor to vitamin A. Shouldn't we support corporations pushing ahead with these vital new products?

Maybe genetic engineering will one day bring these and other benefits to billions of the world's poorest people. But just as pharmaceutical corporations are more interested in finding a cure for male baldness than for malaria—because most of the people who suffer from malaria are poor and would not be able to pay for a vaccine if one were found—so too the corporations developing GM foods focus on farmers in developed countries who can buy their products. Herbicide tolerant crops, for example, are useful only to those who can afford to spray herbicides. Bt cotton has been grown successfully in some developing countries, leading to lower insecticide use and higher profits, but other GM crops have generally been less successful when tried in the developing world. This is not surprising, given that the GM crops are not designed with the needs of developing world farmers in mind.

If genetically modified plants useful for developing countries are to be produced, the investment will have to come from the public sector or private philanthropic foundations, as in the case of golden rice (and the search for a malaria vaccine). That raises the question whether this is the best use of the resources available for helping the world's poor. In his thorough analysis of the potential for increasing food production, Vaclav Smil has concluded that the world can produce enough food for a population of 9 billion people—the point at which some experts expect the world's population to level off around 2050—without using GM foods.[41] There are many well-established ways of improving agricultural practices, while the benefits of GM crops for the poor are unproven. Over several generations, pests will probably develop resistance to the toxins produced by GM crops, and a switch to organic farming could be a surer way to achieve the environmental benefits that are said to be achievable through GM crops. Genetic modification of organisms is a relatively new technique, however, and it is impossible to know in advance what it will or will not achieve. Until we know more, it would be wrong to close the door on it entirely.

From the perspective of consumers like JoAnn Farb, the most immediate ethical question is whether they should buy GM foods. In the developed countries, with their abundant supply of food, ethically motivated consumers may choose to boycott GM foods now, because they pose environmental risks. That does not mean that they will necessarily oppose research into the development of environmentally safe GM foods for poorer countries, where the pay-off from qualities like increased yields and drought resistance could be much greater.

## THE ORGANIC MOVEMENT AND THE ORGANIC FOOD INDUSTRY

The Department of Agriculture's standards may have cleared up confusion about the legal meaning of the word "organic" on food sold in the United States, but they did not end the controversy over what the term should mean. When Julie Guthman was researching *Agrarian Dreams,* a book on organic farming in California, a major organic food broker told her: "If good, nutritious food is going to get out to people—not just yuppie food—you need to have economies of scale."[42]

Mary Ann and JoAnn are part of a swelling minority of Americans for whom these economies of scale make it easier to buy organic products every week. On the other hand, many small organic farmers complain that regulations permit producers to label their food "organic" when in fact they are very far from practicing truly sustainable agriculture. The concerns about the corporate takeover of organic farming that Jim Motavalli had expressed over dinner are widely shared. As organic farming has gained in market share, the food giants have got into it too. Tyson Foods, Coca-Cola, Cadbury-Schweppes, General Mills, ConAgra, Danone, Nestlé, Heinz, Mars, Phillip Morris/Kraft, and Dean Foods have established their own organic brands, or taken over existing ones. These corporations may be complying with the letter of the regulations, as far as the food they label "organic" is concerned, but those for whom organic farming is a way of life, not a business, believe that the corporations will steadily undermine the true philosophy of the movement in order to increase the profitability of organic food.

Elizabeth Henderson is a board member of the Northeast Organic Farming Association, one of the oldest organic farming associations in the country. She was teaching at a university when, at the age of 36, she decided "I wanted to live in a way that was in concert with my beliefs about the environment and community." That was more than 20 years ago. Now she writes books and articles about organic farming while growing 70 different organic vegetables, fruits, and herbs on the 15-acre Peacework Organic Farm near Rochester, New York. The farm operates a "Community Supported Agriculture" scheme: 300 member families, mostly from Rochester, pay the farm a fee that varies with their ability to pay, but is currently between $14 and $20 a week, for 27 weeks a year. The families also help out at the farm, doing a half-day's work three times a season. In return, they get a box of fresh produce delivered to their homes each week. Henderson says she finds it particularly satisfying to farm for a community of people she knows well: "It's not like shipping crates off somewhere, where I never see the customers." That background makes it easy to grasp why she thinks that the label "organic" should say more than that the farm did not use any one of a specified list of synthetic substances. She sees the organic movement as an ethical approach to farming that includes community-building, social justice, respect for farm workers, and openness to the consumers seeing everything that happens at the farm. For her, organic farming is not an industry, but "the alternative to the industrial food system."[43]

## THE DAIRYING DISPUTE

Horizon, the country's largest organic dairy marketer, controlling 55 percent of the market, is at the center of the debate over "corporate organics." Horizon is a subsidiary of Dean Foods, one of the top 20 U.S. food corporations. Its milk carton features a cheerful cow caught in mid-leap. The carton's text tells us that Horizon milk comes from cows who "make milk the natural way, with access to plenty of fresh air, clean water and exercise." It also says, "Happy, healthy cows produce better milk for you and your family." Horizon owns a 4,500-head dairy in Idaho. For an organic dairy, that's huge. When Rebecca Clarren visited it, she found the cows standing in crowded pens, in a stark, arid landscape with no pasture in sight. Her judgment: "The cows don't look that happy."[44]

Aurora Organic Dairy, a subsidiary of a dairying corporation that also produces conventional milk, owns an even larger dairy, with 5,700 cows, on the plains north of Denver. There, most of the cows are kept in pens, outdoors, but at a density far too high to permit grass to grow. They stand on bare earth and are fed on organic grain. They go out to pasture only before they start producing milk and for the brief periods before they give birth to their calves, when they are again not producing milk. The milk is sold mainly to supermarkets like Trader Joe's, to use for their own brand of organic milk.

The Cornucopia Institute, which describes itself as a "a nonprofit group dedicated to the fight for economic justice for the family-scale farming community" and is aptly based in Cornucopia, Wisconsin, has filed a formal complaint with the National Organic Program asking them to investigate both Aurora Organic's Colorado dairy and Horizon's Idaho dairy The regulations say that organic producers must provide "access to pasture for ruminants." The rules recognize some exceptions to these rules, for example for inclement weather or the animal's stage of production, but these exceptions permit only "temporary confinement." Cornucopia contended that these exceptions could not be interpreted as permitting a cow to be denied access to pasture for "the vast majority" of her productive life. Mark Kastel, a senior policy analyst at Cornucopia, told a reporter from the *Chicago Tribune* that "a factory farm is a factory farm" and a dairy farm shouldn't qualify as organic simply because its owners "cram organic feed down the throats of [their] high-producing cattle."

Cornucopia's stance against this kind of "organic" farming received strong support from George Siemon, who went into organic farming in 1977. One of the first to sell organic milk commercially, he joined with six other dairy farmers in 1988 to form an organic dairying cooperative called Organic Valley. The cooperative has grown to nearly 700 farmer-members and is the second largest supplier of organic milk in the nation. Simeon is now its chief executive. Unlike corporate-owned Horizon, however, Organic Valley is still owned by its members, democratically run, and not hierarchical in spirit.

We spoke to Travis Forgues, a Vermont dairy farmer who is an Organic Valley member, just after he'd visited the cooperative's headquarters in LaFarge, Wisconsin. "It's amazing," he told us. "You walk through there and it's not a corporate entity. You can go and talk to George, you can talk to everybody. The farmers actually still have a say in what happens in Organic Valley." In Siemon's view, and probably that of the majority of Organic Valley farmers, there is nothing wrong with the rules for certifying products as organic, except that the U.S. Department of Agriculture isn't being tough enough in enforcing them. "Clearly these people aren't doing pasture as they should be," he told the *Chicago Tribune* reporter. "You can't have these animals on a little piece of land and call it pasture."[45]

The people behind the corporate organic milk giants Horizon and

Aurora Organic Dairy deny that they are selling out the values of the organic movement. Mark Retzloff, president of Aurora Organic Dairy, can trace his career in organic foods back even further than Siemon, to 1968, when, as a student and anti-war protester at the University of Michigan, he helped start an organic cooperative. Now he argues that by making organic milk more affordable, he is responsible for converting more land to organic production. Large organic dairy farms, with 1,000 cows or more, produce 25 to 30 percent of the organic milk produced in the United States. With organic milk sales rising 20 percent each year, those farms are needed to meet the demand. It takes 200,000 acres to grow the organic crops and feed they need—land that otherwise might not have been farmed organically. The cows at Aurora Organic Dairy alone need 50,000 organically farmed acres to grow their feed.[46]

Retzloff's outlook is shared by Steve Demos, a Buddhist who began his business career making tofu in a bathtub and selling it at his tai chi classes in Boulder, Colorado. That business grew into White Wave, a major manufacturer of soy products, including Silk, the best-selling American brand of soymilk. But in 2002, Dean Foods bought White Wave for around $200 million. Although Demos at first tried to block the takeover, Dean Foods persuaded him to stay on when that failed, and, ironically for someone whose life mission appeared to have been to replace dairy products with soy products, Demos also became head of that other Dean Foods subsidiary, Horizon. Demos defends his corporate approach by saying that unless the organic movement changes, "You'll have an elitist industry selling niche products at three times what the average person can afford."[47] Undeniably, Horizon's financial resources have enabled it to push organic milk into places like local supermarkets and Starbucks and even school programs, where it was not found previously.[48]

In fairness to Horizon, we need to point out that in addition to getting milk from its Idaho mega-dairy and some other very large organic dairies, it also buys from 300 family farmers. Among them are Rodney and Judith Martin, a Mennonite couple who farm 260 grassy acres, near Oxford, Pennsylvania. Formerly conventional dairy farmers, they are now enthusiastic advocates of organic dairying. When we visited, we saw rolling green fields with shade trees around two sides. Black-and-white cows, about 70 of them, their heads down, were eating the fresh grass. Others were lying down chewing their cud. There were maybe a dozen calves in the field too, some near their mothers, others lying down. It was the way a dairy farm is supposed to look, but few still do.

"Why cut forage and store it and give it to the cow," Rodney Martin asked rhetorically, "when the cow loves to go out there and cut her own forage?" The cows can come into a barn if they wish, but they prefer to be outside for at least three-quarters of the year, and sometimes will stay out even in cold, wet weather. They are only prevented from going out for about a dozen days a year, when the weather gets really nasty and Rodney Martin isn't prepared to leave it to them to make the decision to go out or stay in. He says he gets a lower milk yield than confinement dairies, but, he told us, he gets "a healthier cow, healthier product, happier environment, happier people—and financially it's every bit as good or better." The Martins let their calves run with their mothers until they are around 10 weeks old, which is very different from the usual practice in the dairy industry, but even at that age, separation will be distressing for both the calves and the cows.

While Jim Motavalli would take the side of the independent organic farmers in this dispute, Mary Ann Masarech, with her own background in the corporate world and the many demands on her time, is much more sympathetic to the corporate stance. She's a pragmatist, not a purist. "If you want to make change," she said, "you've got to make it easier for people." The big corporate organic farmers are making it easier for people to get organic produce in their local supermarkets.

In March 2005, a federal advisory panel on organic standards recommended that the rules specify that organic milk must come from cows allowed to graze on pasture for at least 120 days per year. As of this writing, this recommendation is subject to approval from the Department of Agriculture and could take some years to come into effect.[49]

## The Ethics of Organics

Marion Nestle has worked as a nutritionist for more than 20 years. During that period, she served as a nutrition policy adviser to the U.S. Department of Health and Human Services and was a member of committees advising the Department of Agriculture and the Food and Drug Administration. After leaving government, she became professor and chair of the Department of Nutrition and Food Studies at New York

University. She wrote *Food Politics,* a stinging critique of the influence of the food industry on government policy. Although her interest in organic foods developed only recently, she is now an enthusiastic purchaser and advocate. Despite her background in nutrition—some scientific studies show that organic foods have slightly higher levels of some nutrients than conventional foods—for Nestle it is ethics, rather than personal health, that provides the important reasons for buying them. The true value of organic foods, she argues, comes from what they do for farm workers, who have lower pesticide exposure; for animals, who are more humanely treated; for soils, which are enriched and conserved; for water supplies, which have less fertilizer runoff; and for other environmental causes. Nestle says she has been asked countless times whether the seal saying a product is organic really means anything, and her answer is that it does. In her experience, inspectors and producers alike work hard to preserve the trust of consumers.[50]

There are really only two serious objections to these ethical reasons for consumers in developed countries to buy organic food. The first is that because conventional farming has a higher yield for a given area of land, it allows us to leave more land for other purposes, including preserving natural areas and wildlife.[51] The second is that organic food is too expensive for the poor to buy.

It is generally true—but not in all cases—that organic farming methods produce lower yields than conventional methods, but the difference is small, and as we have seen, a Rodale Institute study found that in the long-term, organic farm yields were as high as conventional farms. But even if it were true that to generate comparable yields, we must choose between conventionally farming 85 acres, leaving 15 for wildlife, and farming 100 acres organically, organic farming would still be more environmentally responsible, because of its lower energy use, less pesticide runoff, healthier soils, reduced erosion, and more biodiversity on the farm itself.

Organic food costs more partly because, as we have seen throughout this book, intensive industrial agriculture leaves others to pay the hidden costs of cheap production—the neighbors who can no longer enjoy being outside in their yard; the children who cannot safely swim in local streams; the farm workers who get ill from the pesticides they apply; the confined animals denied all semblance of a life that is normal and suitable for their species; the fish who die in the polluted streams and coastal

# 15

# IS IT UNETHICAL TO RAISE CHILDREN VEGAN?

Shortly after we visited the Farbs and their apparently thriving children, we were surprised to see headlines proclaiming that vegan diets are "bad for children." The articles quoted Professor Lindsay Allen of the U.S. Agricultural Research Service saying, "There's absolutely no question that it's unethical for parents to bring up their children as strict vegans." Withholding animal products from children, Allen was reported as saying, can do them "irreparable physical and mental harm." In support of this claim, the articles mentioned a study that Allen had done on children in Kenya. One report said that the study was partially funded by the National Cattlemen's Beef Association.

The reports drew a storm of angry comments from vegans and an immediate rebuttal from Sir Paul McCartney, the former Beatle and long-time vegetarian, who described Professor Allen's views as "rubbish" and said that the studies on which the comments were based "are engineered by livestock people who have seen sales fall off." He added: "Vegetarianism has been a good thing for me and my children, who are no shorter than other children." In response, Allen was quoted as saying that the study was done for the U.S. Agency for International Development, not for a meat company, but she wasn't going to lose sleep over the fact that Sir Paul was upset.[1]

The American Academy of Pediatrics has stated that vegan diets can promote normal infant growth. The American Dietetic Association says that "Well-planned vegan and other types of vegetarian diets are appropriate for all stages of the life cycle, including during pregnancy, lactation, infancy, childhood, and adolescence," adding that vegetarians have lower rates of heart disease, type 2 diabetes, hypertension, and prostate and colon cancer.[2] Many vegetarians and vegans are adamant that vegetarian and vegan diets are not only nutritionally adequate, but actually healthier than conventional diets; they counter that parents who raise their children on the conventional Western diet, heavy on animal products, are harming their children. After all, vegetarian diets are usually lower in fat and higher in fiber than conventional diets, and experts tell us fat is bad and fiber is good. Vegan diets are particularly low in saturated fats, which generally come from animals and are a major cause of heart disease. Furthermore, vegan food won't pass on infections like the rare but fatal variant of Creutzfeld-Jakob disease caused by eating beef from animals infected with mad cow disease, and vegans are much less likely to get the more common but still dangerous salmonella and *E. coli* infections.

The advocates of vegetarian diets often point to exceptionally long-lived vegetarian or near-vegetarian peoples, like the Hunzas of northwest Pakistan, or the people of the Vilcabamba valley in Ecuador. But questions have been raised about the accuracy of the birth records of these rural peoples, and even if they do live significantly longer than the rest of us, it might be because of the high altitude at which they live, or their healthy outdoor lifestyle, or their genes, or something in the water. Of greater scientific significance are the large long-term comparative studies of vegetarians and non-vegetarians that have recently been carried out, over several decades, in developed countries. Studies of Seventh Day Adventists, many of whom are vegetarian, have shown them to live longer than the general population. This could, however, be related to lower rates of smoking or alcohol use. It is therefore important to try to control for such variables.

In an article published in the *American Journal of Clinical Nutrition* in 2003, Pramil Singh, Joan Sabaté, and Gary Fraser examined six major comparative studies carried out in Europe and North America, and cautiously concluded that they "raise the possibility that a lifestyle pattern that includes a very low meat intake is associated with greater longevity." Within the definition of "very low meat intake" the authors included eating no meat at all. One of the six studies found no association at all between very low meat intake and longevity, while the other five found that people on a very low meat diet lived longer. The difference was statistically significant in four of these five studies, and two of them suggested a 3.6-year increase in life-expectancy for those on the low meat diet.[3]

No wonder, then, that vegan and vegetarian Web sites were soon full of vituperative comments about Allen's reported remarks. One story posted on a prominent vegetarian Web site began with the headline "National Cattlemen's Beef Association pays for Sadistic Anti-Vegan 'Study'" and included this sentence: "Like Nazis experimenting on captives, the cattle industry manipulated very slightly the diets of starving African children—not to benefit the children but to try to produce some 'scientific finding' which justifies meat-eating."[4]

We were puzzled. Why was Allen rejecting well-accepted medical and scientific views on this issue? More troubling still, how could a study of children in Kenya be the basis for so sweeping a condemnation of vegan diets when obviously these children wouldn't have access to the same wide range of vegan foods as children in developed countries? We also knew that Allen is a respected scientist who has served as President of the American Society of Nutritional Sciences and as an advisor to the World Health Organization. We weren't ready to assume that she could easily be bought by the cattle industry. We decided to get her side of the story. It turned out to be yet another illustration of the old adage that you can't always believe what you read in the newspapers.

Allen has worked in the field of nutrition in developing countries for 25 years. At the 2005 annual meeting of the American Association for the Advancement of Science, she reported on a study in which some Kenyan children were given additional animal-based foods, while others received a matching amount of additional calories from plant foods and a control group continued on their normal food intake. The children receiving animal foods were themselves divided into two groups, some of them getting meat, and others a cup of milk. The outcome was that the children receiving additional animal foods—both those receiving meat and those getting milk—did significantly better, on a variety of health and growth measures, than the group that received additional plant foods, or the control group. None of the children were "starving" and none of them were harmed by the study. Allen denies that her own research was paid for by the National Cattlemen's Beef Association, but the study on which she reported, and of which she was a co-author, did get some funding from that organization, and the publication of a special supplement to the Journal of Nutrition that Allen edited, which included reports on this and other studies of the use of foods from animal sources to improve nutrition in developing countries, received funding from, among others, the agribusiness conglomerate Land O'Lakes Inc.[5]

After the panel, a BBC reporter came up to Allen and asked her if the study she had described was relevant to vegan children in developed countries. At this point in telling us her story, Allen paused for a moment before saying: "Like a fool, I said, 'Yes.'" But, Allen continued, she added a vital caveat. "I said it was unethical for people to bring up their children as vegans, *unless they take great care to know what they're doing.*" The BBC left off the caveat—even though Allen insists she said it twice—and her unqualified comment then got picked up everywhere. Allen also insisted that she had never said she didn't care about Sir Paul McCartney's opinion.

We asked Allen what she really does think about vegan parents bringing up their children without any animal products (except for breast milk, which almost all vegans favor for infants). "I'm not against veganism," she said. "I'm against people who, often because of an animal rights ideology, don't take the trouble to learn about what they should be eating. People come out with self-righteous attitudes and lots of pure malarkey about how you can get vitamin $B_{12}$ from plants or from the soil. I don't believe that."

Early vegans sometimes had health problems because they were not aware that their diet did not provide them with vitamin $B_{12}$. That may be the origin of the widespread myth that humans cannot be well-nourished without animal products. In developed countries, most vegans now take a $B_{12}$ supplement or eat foods fortified with $B_{12}$ and enjoy good health. Allen's focus, though, is on nutrition in developing countries. She tells us about success stories of schemes to improve the health of children in poor countries by assisting women with keeping small animals. Hens, for example, can be kept on land that can't be used for anything else, and their eggs can help to nourish families. "What makes me angry," she says, "is when people say that we shouldn't be giving animal foods to people in developing countries, where there are no $B_{12}$ supplements or Linda McCartney–prepared vegetarian meals."

What Allen refers to as an "animal rights ideology" may, of course, be a carefully thought out ethical position—as it is for the Farbs. But when JoAnn became pregnant, she did her homework, reading medical and scientific sources on vegan pregnancies and children raised as vegans. You don't need to go directly to the scientific journals for this purpose. There are several vegan advocacy groups that provide scientifically sound advice. One of the best is Vegan Outreach, which states that it promotes a vegan diet to lessen the suffering of animals, but is careful to provide nutritional information to all who pick up its literature or visit its Web site.[6] Allen does not disagree with JoAnn's view of the science—that it is safe to be a vegan while pregnant and to raise your children as vegans—so long as you are careful in what you and your children eat.

## VEGAN HEALTH

The first question most people ask vegans and vegetarians is, "Where do you get your protein?" The prevalence of that question is a hangover from an earlier period, starting in the 1930s, when nutritionists believed we needed a high level of protein in our diet, and that a protein-deficient diet was a major cause of malnutrition. A lack of protein was a problem among some people eating little else but the tropical root crop cassava—eaten in the West mainly in the form of tapioca—which contains less protein than any other major staple food. But other malnourished people thought to be suffering from protein deficiency were in fact simply not getting enough calories. The idea that we need high levels of protein was disproven in the 1970s, and health authorities reduced recommended protein intakes to about a third of what had been thought to be required. These lower levels mean that for adults eating sufficient calories, even a diet consisting exclusively of bread, pasta, rice, or potatoes will provide adequate protein. Children and pregnant and breast-feeding women need more protein, but that is easily provided by adding beans, peas, or lentils to basic staples like bread, pasta, or rice.

JoAnn said that they ate plenty of nuts and seeds and that when pregnant, she was careful to combine different forms of protein at each meal, like rice and lentils, or beans with a tortilla. She may have been more careful than she really needed to be. The American Dietetic Association states: "Plant protein can meet requirements when a variety of plant foods is consumed and energy needs are met. Research indicates that an assortment of plant foods eaten over the course of a day can provide all essential amino acids and ensure adequate nitrogen retention and use in healthy adults, thus complementary proteins do not need to be consumed at the same meal."[7]

After protein, the nutrient that concerns most people when going on a vegetarian or vegan diet is iron. Many plant foods are rich in iron, including soybeans, pumpkin seeds, dried apricots, pinto beans, spinach, and raisins. Iron from plant foods may not, however, be as easily absorbed as iron in meat—this is particularly the case with the iron in spinach. Taking vitamin C with iron-rich foods makes the iron more available, whereas calcium supplements, coffee, and tea inhibit the absorption of iron. In some countries staple foods are fortified with iron—as wheat flour is in the United States. Most vegetarians and vegans in developed countries do not need to take iron supplements. What a purely plant-based diet will not provide—Allen is quite right about this—is vitamin $B_{12}$. Responsible vegan organizations recommend that vegans and near-vegans take a $B_{12}$ supplement.[8] Another vitamin that is not abundant in a vegan diet is vitamin D, but most people get enough vitamin D by going out in the sun. Dark-skinned people, however, do not absorb as much vitamin D from the sun, and of course those who cover their bodies for religious reasons, as many Moslem women do, cannot get vitamin D from the sun. In high-latitude winters, a vitamin D supplement may also be advisable. American margarines, including vegan options, are fortified with vitamin D.

The Farb family does all the right things. They eat a varied plant-based diet, with no junk food or other "empty calories." They also take a $B_{12}$ supplement and, in winter, a multi-vitamin supplement that contains vitamin D. When Allen's real views are disentangled from the media's distortions, she clearly isn't saying that the Farbs are doing anything wrong in raising Sarina and Samantha as vegans. But to say that a vegan diet is good for families with the choices available to the Farbs isn't to say that it is good for everyone.

The work of Allen and her colleagues suggests that in rural areas where $B_{12}$ supplements are unavailable, children will develop better if they have access to some animal foods. They would also do better if they could receive $B_{12}$ supplements, perhaps as part of a fortified common food. Since we need only tiny amounts of $B_{12}$, this can be done very cheaply. As long as there is a basic food that is commercially processed, can be fortified, and is eaten by practically everyone, fortification will ensure an adequate $B_{12}$ intake for nearly the entire population. In Israel, after studies showed that even meat eaters were not getting enough of the vitamin, flour was fortified by law with $B_{12}$. But it would not be so easy to reach isolated communities that produce most of their own food locally. A study of which Allen is a co-author indicates that adding animal foods to the diet of Kenyan schoolchildren did not lead to any detectable difference in the levels of any micronutrient other than $B_{12}$, although Allen told us that the prevalence of malaria and parasites in the children made it difficult or impossible to measure the levels of some micronutrients, including iron.[9]

Some people seem to think that if one cannot say that everyone in the world ought to become vegan, then there can be no moral obligation on any of us to do so. In his otherwise useful book *So Shall We Reap,* Colin Tudge claims that unless it is the case that everyone in the world ought to become vegan, there can be no moral obligation on any of us to do so. He points out that for many people living in difficult environments at high latitudes or in semi-deserts, vegetarianism is not a viable option. He then draws on the 18th-century German philosopher Immanuel Kant as an authority for the view that "no ethical principle is really acceptable unless it could in principle be recommended to the whole world." He concludes that vegetarianism and veganism fail this test and therefore cannot be ethical principles.[10]

Tudge is referring to Kant's first formulation of the so-called "categorical imperative," or supreme moral law, which is more accurately translated as "I ought never to act except in such a way that I can also will that my maxim should become a universal law."[11] But to understand Kant as Tudge suggests turns his moral law into nonsense. It would mean, for example, that it would be unethical to become a teacher, because if everyone in the world became a teacher, there would be no farmers to grow food for them. It is perfectly possible to restrict the application of moral principles to specific contexts, and we do this all the time. We tell people to keep their promises, but not when the only way to save the life of a road accident victim is to break your solemn promise not to be late for the start of the school concert in which your daughter is playing the violin. Similarly, we could say, "Be vegan, unless the circumstances in which you live prevent you nourishing yourself properly from plants alone." We doubt that there is anything in Kant's philosophy, properly understood, that prevents that being a valid ethical principle—but if there is, so much the worse for Kant.

What about those animal rights groups that make Allen so angry by saying we shouldn't be giving animal food to people in developing countries? We spoke to the leaders of two organizations that have campaigned against the spread of animal production in the developing world: Joyce D'Silva of Compassion in World Farming and Andrew Tyler of Animal Aid. They both made it clear that what they were really opposed to was the introduction of the factory farm mode of animal production. Tyler told us: "Ultimately, my objection is to the commercial forces that are seeking to persuade people of the poor world that their best nutritional interests are served by buying into modern, high-throughput farmed animal production processes. With that comes an addiction to high capital input systems, a loss of control over the means of production, bad health, and a nightmare animal welfare scenario." D'Silva told us of a recent trip to China, where she had seen a totally unsuitable, Western financed, intensive dairy farm that, in addition to being an inefficient form of food production and having poor animal welfare, was causing severe local pollution. China is importing more and more grain and soy in order to feed its factory-farm animals. It is this trend towards Western-style factory farming, rather than "a few chickens running around a village" that Compassion in World Farming is trying to stop.[12]

Meanwhile, further evidence of the ability of a vegan diet to provide all the energy and stamina anyone needs was provided by Scott Jurek's victory in the 2005 Badwater Ultramarathon, one of the toughest events in ultra-long-distance running. Jurek, a vegan, shattered by more than 30 minutes a course record that some thought unbreakable, finishing a full two hours ahead of his nearest rival. Starting below sea level in Death Valley, he ran 135 miles, some of it in 115-degree heat, to finish over 8,000 feet up on the slopes of Mt Whitney—and did it in 24 hours, 36 minutes, and 8 seconds. Along the way, he ate vegan energy bars, potatoes, rice balls, and soy protein drinks. Jurek has plenty of predecessors, among them Carl Lewis, who won nine Olympic track and field gold medals between 1984 and 1996. Lewis became a vegan in 1990 and has written that "my best year of track competition was the first year I ate a vegan diet."[13]

JoAnn Farb wouldn't keep up with either Lewis or Jurek, but she took up jogging at the age of 43 and found that after a few weeks, she could run three miles without stopping. The Farb family has enjoyed excellent health on their vegan diet. Neither Sarina nor Samantha has ever been in a hospital, broken a bone, or even taken antibiotics.

# 16

# ARE VEGANS BETTER FOR THE ENVIRONMENT?

Some people think that factory farming is necessary to feed the growing population of our planet. The truth, however, is the reverse. No matter how efficient intensive pork, beef, chicken, egg, and milk production become, in the narrow sense of producing more meat, eggs, or milk for each pound of grain we feed the animals, raising animals on grain remains wasteful. Far from increasing the total amount of food available for human consumption, it reduces it.

A Concentrated Animal Feeding Operation, or CAFO, is, as the name implies, an operation in which we concentrate the animals together and then feed them. Unlike cattle on pasture, they don't feed themselves. There lies the fundamental environmental flaw. Every CAFO relies on cropland, on which the food the animals eat is grown. (Aquaculture for carnivorous fish like salmon has no need of cropland, but the basic principle is the same—they don't catch fish, fish must be caught for them.) Because the animals, even when confined, use much of the nutritional value of their food to move, keep warm, and form bone and other inedible parts of their bodies, the entire operation is an inefficient way of feeding humans. It places greater demands on the environment in terms of land, energy, and water than other forms of farming. It would be more efficient to use the cropland to grow food for humans to eat.

Cattle in feedlots mostly eat grains—in the U.S., corn, but in other countries it may be wheat or another grain. This is what Frances Moore Lappé famously called "a protein factory in reverse"—meaning that you start out with a large amount of protein, channel it through cattle, and end up with a much smaller amount.[1] In her 1971 classic *Diet for a Small Planet,* she calculated that it took 21 pounds of grain to produce one pound of beef, and that an acre of land devoted to cereals could produce five times as much protein as an acre devoted to meat production. Since then, beef producers have improved their efficiency, but when we take into account the fact that only about half the weight of a steer is boneless beef, 13 pounds of grain are required to produce that single pound of beef.[2] With pigs, it takes about six pounds of grain to produce one pound of boneless pork.[3] But even these figures are flattering to meat production, because a pound of meat contains much more water than a pound of grain does.

Raising chickens is less inefficient. According to the U.S. National Chicken Council, it takes just two pounds of feed to produce one pound of chicken, but this is a live-weight figure. After slaughter, when blood, feathers, and internal organs have been removed, a 5-pound chicken won't produce much more than 3 pounds of meat.[4] That puts the grain-to-meat conversion ratio back up over 3 to 1, including bones and water. So the National Chicken Council's own figures prove that, even with the most efficient form of intensive meat production, if we really want to feed ourselves efficiently, we'll do much better to eat the grain ourselves than to feed it to the chickens. If it is protein, rather than simply calories, we are after, we'll do better still growing soybeans. Although in the past some nutritionists claimed that animal protein is higher in "quality"—that is, in the balance of amino acids—than plant protein, we now know that there are no significant differences in the quality of protein between soybeans and meat.[5]

Experts on agricultural efficiency agree with this conclusion. According to Vaclav Smil's *Feeding the World,* producing milk, meat, and eggs based on feeding the animals crops we grow for that purpose "inevitably" leads to less food than we could produce from the land on which the crops were grown. This holds, Smil writes, whether we are talking about the energy the food provides or the amount of protein we obtain from it. Hence, Smil calculates, it is simply not possible for everyone in the world to eat as much meat as people in the affluent world now eat, because to produce that amount of meat would, in the absence of some unforeseen advances in bioengineering, require 67 percent more agricultural land than the world possesses.[6]

Other environmental concerns often raised about rearing animals are the clearing of forests, especially rainforests, and the quantity of water used. Sarina Farb mentioned both of these when we visited her family. After first saying that her vegan diet is "good for the animals," she added: "Thousands of trees are being cut down to raise cattle. The rainforests are being cut down. And we're using lots and lots of water just to grow one pound of beef."

We've all read about the alarming rate at which forests are disappearing. The Amazon rainforest, for example, is still being cleared at an annual rate of 25,000 square kilometers, or 6 million acres, to graze cattle and grow soybeans to feed to animals.[7] That translates to another 11 acres felled every minute. But what does that have to do with American beef consumption? If the Farbs were to go to McDonald's for their next meal, they would not find any beef that came from recently cleared rainforest. As we saw in Part I, McDonald's has a long-standing policy against purchasing such meat, and in the McLibel trial, Mr Justice Bell found that Helen Steel and David Morris's claims about McDonald's contribution to the destruction of rainforests were defamatory.

Nevertheless, Gaverick Matheny and Kai Chan, researchers from the fields of agricultural economics and conservation biology, respectively, have pointed out that, thanks to the reduction of trade barriers, the world is, to a far greater extent than previously, now a single market. The global increase in meat consumption is causing forests in other countries to be cleared to grow grains and soybeans for feeding to animals. Thus our own meat consumption contributes indirectly to deforestation abroad, with a consequent loss of biodiversity.[8] We could be eating products from animals raised in the U.S., fed on Brazilian soybeans that were grown on land cleared from rainforest. Also, when we buy meat, eggs, and milk that were produced using land that might otherwise have grown crops for export, additional land might have been cleared abroad as a consequence. Even if we eat meat produced in the U.S. from animals fed soybeans grown in the U.S., the export of those soybeans could have met the demand that was instead met by soybeans grown on the cleared rainforest land.

So the knowledge that we are not ourselves eating "rainforest beef" from cattle grazed on cleared rainforest is not sufficient to tell us whether our food choices are likely to influence the rate of deforestation in the Amazon or elsewhere. Of course, if we consume soybeans, soy milk, tofu, or other soy products, we are also linked into the clearing of forests. But it would be a mistake to think that since we cannot isolate our food from the clearing of the rainforests, it doesn't matter whether we eat beef or soy. Most of the soy grown in Brazil is exported for use in animal feed. Since feeding soy to animals yields only a fraction of the original food value, an animal-based diet will make a greater contribution to clearing rainforests than a diet that is based on eating grain or soybeans directly.

## WATER CONSUMPTION

Sarina Farb's other environmental reason for being a vegan is "we're using lots and lots of water just to grow one pound of beef." Water use has become part of the standard armory of arguments against eating meat. Agriculture consumes more than 70 percent of the world's developed freshwater supplies. Without water, we cannot survive, and in many parts of the world, water shortages are becoming critical. So much water is taken from rivers for irrigation that there is insufficient flow to maintain the health of many key U.S. rivers, and underground aquifers are being depleted. In the United States, one giant natural storage basin—the Ogallala Aquifer, which stretches from South Dakota across eight states to Texas—supplies water to 65 percent of the irrigated acreage in the nation. It took thousands of years to fill, but it is now being pumped out much faster than it can replenish itself, and water levels are falling. Farming that relies on this water has been likened to a mining operation rather than a sustainable form of agriculture. It exploits a finite resource, makes a profit from it, and when the resource is gone, leaves behind a landscape that can never be restored to what it once was. Hence if a single pound of beef really does require "lots and lots of water," then raising beef cattle is, in many parts of the U.S. and in some other countries too, a profligate use of a limited and precious resource.

In 1981, when much of the U.S. suffered from drought, a *Newsweek* special report on "The Browning of America" asserted: "The water that goes into a 1,000-pound steer would float a destroyer."[9] That factoid was later popularized by John Robbins, the heir to the Baskin-Robbins fortune, who turned his back on the family's ice-cream business and advocated a vegan diet in his widely read *Diet for a New America*, published in 1987. From there, the "enough water to float a destroyer" line has taken on a life of its own. Is it really true? We decided to check.

The *Newsweek* article said it may take 3,500 gallons of water to produce a steak, but provided no source for the figure. Since then, many other estimates have been published, ranging from 441 gallons to about 12,000 gallons per pound of beef. The lowest figure comes from a study by J. L. Beckett and J. W. Oltjen of the University of California. Davis, supported by the California Beef Council.[10] The highest figure is from a 1997 article by a team led by David Pimentel, a widely-published ecologist at Cornell University.[11] Pimentel's article gave little information on how the figure was reached, so we contacted him. He sent us a more recent article in which he and his co-authors have reduced their estimate to 5,152 gallons per pound.[12] We also found a report by A. K. Chapagain and A. Y. Hoekstra, published in 2004 by the UNESCO-IHE Institute for Water Education, in Delft, Netherlands, that gives a global average figure for beef of 1,860 gallons of water per pound, with an average of 1,584 gallons for beef produced in the United States and 2051 gallons for Australian beef.[13]

We were surprised by the wide variation in these studies, all apparently by reputable scientists. However, the major reason for the differences was easy to spot. The studies include in their calculations very different proportions—ranging from 0 to 100 percent—of the water that falls as rain on the rangeland on which cattle graze. Including all the rain seems mistaken. Grazing cattle do not use all of the rain that falls on the land on which they are grazing. Chapagain and Hoekstra, whose study is the most sophisticated we have seen, calculate how much the cattle eat, and then include only the water taken up by that quantity of plants, not the water that runs off into streams, soaks into the ground, or is used by plants not eaten by the cattle. That is more defensible. As we have seen, Chapagain and Hoekstra's estimate of the amount of water it takes to produce a pound of beef is less than one-third of the estimates at the higher end of the range. On their figures, the U.S. beef industry "only" uses 792,000 gallons of water to produce a 1,000-pound steer, not nearly enough to float a modern destroyer, although naval buffs could point out that World War II destroyers were much smaller, and 792,000 gallons would float some of them.[14]

If our purpose is to calculate the environmental costs of producing beef, we should be particularly interested in water that comes from scarce or nonrenewable sources. For example, if water is drawn from underground aquifers at a rate faster than it is being replenished, that is clearly something to be concerned about. We should also focus on water diverted from a river to irrigate pastures, when that water could have been used to grow crops, to provide drinking water for cities, or just left in the river to replenish wetlands that provide habitat for many wild animals and birds.

At the low point of the range, Beckett and Oltjen set out to measure only "water that is developed and is diverted from possible use by humans." They did not include "natural precipitation," whether it fell on rangeland or on crops. This is to err in the opposite direction from those studies that include all rainfall. The primary reason for not counting rainwater falling on grazing land is that we can't collect this water or use it for other purposes, like growing corn or tomatoes—most rangeland is unsuitable for farming. However, in the absence of cattle, the rain might produce natural vegetation that would make the area more attractive for recreational purposes. Furthermore, the rain falling on cornfields that are a source of food for cattle could be used for commercial farming, since that land could instead be used to grow food directly for humans. Also, if water is a scarce resource, then crops grown with natural rainfall can replace crops grown on irrigated land. So it seems arbitrary to exclude rain that falls on crops fed to cattle, while including irrigation water used on crops fed to cattle. Excluding the rainfall required to grow corn for cattle greatly lowers the estimate of the amount of water used in beef production, because most corn grown in the U.S. is not irrigated.

To say that we are using "lots and lots of water just to grow one pound of beef" is, then, a much more complex claim than Sarina Farb could possibly have imagined. In our view, Chapagain and Hoekstra's study is the best one to use, although it simply estimates the aggregate water consumed in producing a pound of beef and does not purport to measure the amount of water that we could have used for other purposes if the relevant land had not been used to produce beef. Even on Chapagain and Hoekstra's estimates, however, beef uses much more water than other food staples. Producing a pound of hamburger beef will take 12 times as much water as a pound of bread, 64 times as much as a pound of potatoes, and 86 times as much as a pound of tomatoes.[15] Even after allowance is made for the calories per pound supplied by different foods, from the standpoint of water conservation beef isn't a good choice. Bread delivers roughly the same calorie count as hamburger beef, for one-twelfth of the water usage.

Still, this isn't a powerful argument for eating vegan. As far as water usage is concerned, no vegan willing to consume a 200-gram bag of potato chips is in any position to criticize someone who eats an egg, because, according to Chapagain and Hoekstra, the chips took more water to produce. Milk and apple juice, they calculate, use virtually identical amounts of water to produce. In the U.S., chicken meat requires only about 20 percent more water per pound than soybeans or rice. Figures vary in different countries because different production methods are used, but of all the basic foods we eat, beef is the only one that really stands out from other foods as requiring substantially more water.

## DEGRADING THE LAND

> The rancher (with a few honorable exceptions) is a man who strings barbed wire all over the range; drills wells and bulldozes stockponds; drives off elk and antelope and bighorn sheep; poisons coyotes and prairie dogs; shoots eagles, bears, and cougars on sight; supplants the native grasses with tumbleweed, snakeweed, povertyweed, cowshit, anthills, mud, dust and flies. And then leans back and grins at the TV cameras and talks about how much he loves the American West.[16]

So said Edward Abbey, the author of *The Monkey Wrench Gang*, a novel that inspired some environmental activists to engage in economic sabotage to protect the environment. According to the World Resources Institute, overgrazing is the largest single cause of land degradation, worldwide.[17] Much of this degradation occurs in the semi-arid areas used for cattle and sheep grazing in countries like the United States and Australia.

Ranchers tend to be politically conservative, and conservatives are generally loud in their support of the market. Curiously, however, grazing rights over public land in the United States are not sold by auction or competitive tender. Instead, politics and the disproportionate political clout that the U.S. constitution gives to states with small populations, has led to ridiculously low grazing fees—in some parts of the country, just a few cents per acre—and additional government support for ranchers. In the United States, the fees received by the government don't even cover the expense of regulating and supporting the cattle grazing. Hence the apt title of one influential book on the environmental problems caused by the beef industry in the United States: *Welfare Ranching*.[18]

A staggering expanse of public land is leased for grazing in the United States—300 million acres, as much as all the eastern seaboard states, from Maine to Florida, with Missouri added.[19] Nearly 90 percent of it is federal land, controlled by the Bureau of Land Management or the U.S. Forest Service. Cattle are heavy animals with hard hooves, big appetites, and a digestive system that produces a lot of manure. Turned loose in fragile, semi-arid environments, they can soon devastate a landscape that has not evolved to cope with them. During dry periods they eat almost everything that grows. Without plants to bind the soil, wind erosion carries away much of the topsoil. When heavy rain falls again, the water sweeps away the bare soil and cuts deep gullies into the landscape. Cattle hooves compact soil and damage river banks, and manure pollutes streams and breeds flies.

When cattle have been kept out of an area for a decade or two, the difference can be dramatic. Twenty years ago, the area that is now the San Pedro Riparian Conservation Area in southeastern Arizona was barren and desolate, the landscape mostly dry earth with sparse vegetation eaten down by 10,000 cattle. The cattle trampled the banks of the river, causing it to become wide and shallow. In 1988, Congress set the area aside for conservation and cattle were excluded from most of it. Today the San Pedro River again runs narrow and deep, native grasses and bushes have grown up, and there are groves of cottonwood and willow. Tens of thousands of birdwatchers come to the area each year to see some of the 350 species of birds that live there or migrate through it. There are also 81 species of mammals and 40 species of reptiles and amphibians.[20]

There may be a few ranchers who appreciate biodiversity and the beauty of a natural ecosystem, but in their farming practices most of them view nature in terms of increased or decreased productivity of the land for their grazing animals. Predators will be killed without question. So too will animals that eat the grass "intended" for the rancher's cattle or sheep. The Australian government, under pressure from the graziers, issues permits for the killing of 4 or 5 million kangaroos every year.[21] In the United States, ranchers have virtually exterminated prairie dogs, social rodents who live in large underground "towns" and eat grass. Over the past century the U.S. government has, at the behest of ranchers, sponsored vast prairie dog poisoning programs. Poisoning still continues today, although necessarily on a smaller scale, since the prairie dog population is now only about 2 percent of what it once was. U.S. Forest Service Biologist Dan Uresk has done studies showing that while prairie dogs do reduce the amount of forage available to cattle—by approximately 4 to 7 percent—the cost of poisoning them outweighs the gains made by eliminating them. However, because the government pays for the poisoning, Uresk says, "Ranchers think: 'Why not poison them if it's free?'"[22] Getting rid of prairie dogs creates flow-on effects for other animals. There are, Uresk has shown, roughly twice as many species—owls, hawks, eagles, plovers, foxes, ferrets, and small rodents—around prairie dog towns as there are in similar areas without them.

The U.S. Department of Agriculture has a "Wildlife Services" division that used to be called "Animal Damage Control." Environmentalists used to joke that the initials stood for "All Dead Critters" or "Aid to Dependent Cowboys," but the name did give an honest indication of what the division does. In 2004, Wildlife Services killed by poisoning, shooting, or trapping, 2,767,152 wild animals, including badgers, beavers, bears, blackbirds, coyotes, doves, finches, foxes, geese, marmots, opossums, prairie dogs, raccoons, ravens, skunks, squirrels, starlings, and wolves.[23] Many of these animals will have died lingering deaths from poisoning. Perhaps worse still, was the fate of 6,485 coyotes caught in notorious steel-jawed leg-hold traps. Sometimes coyotes are trapped for days in blazing sun without water. In their desperation to escape, some animals have been known to chew off their foot.

There are, as even Edward Abbey acknowledged, "honorable" ranchers who care for the land, grazing their animals in a sustainable manner and happy to share their land with the wild animals who belong there. We described one such beef producer in Chapter 4, and there are many others. For the purchaser of beef or lamb, however, it usually isn't possible to find out how well the rancher on whose land the animal grazed cared for the land. Hence, environmental concerns raise an ugly question-mark over the ethics of eating meat from grazing animals.

More generally, and despite the availability of sustainably produced animal products, vegans are right to say that their diet is far more environmentally-friendly than the standard American diet. Gidon Eshel and Pamela Martin, of the University of Chicago, studied the greenhouse gases emitted by the production of animal products, and concluded that the typical U.S. diet, about 28 percent of which comes from animal sources, generates the equivalent of nearly 1.5 tons more carbon dioxide per person per year than a vegan diet with the same number of calories. By comparison, an average driver switching from a typical American car to one of the more fuel-efficient hybrids would save 1 ton of carbon dioxide per year—making the switch to a vegan diet a more effective way of reducing one's contribution to climate change.[24] (Though it would, of course, be better still to do both.)

The editors of *World Watch Magazine*, the journal of the Washington, DC–based World Watch Institute that follows global environmental issues, have noted that the "seemingly small issue of individual consumption" of meat has now become central to discussions of sustainability. This is because, they wrote, "as environmental science has advanced, it has become apparent that the human appetite for animal flesh is a driving force behind virtually every major category of environmental damage now threatening the human future—deforestation, erosion, fresh water scarcity, air and water pollution, climate change, biodiversity loss, social injustice, the destabilization of communities and the spread of disease."[25]

# 17

# THE ETHICS OF EATING MEAT

Many people, like Jake Hillard and Lee Nierstheimer, eat whatever meat takes their fancy at the supermarket or in fast food restaurants. Some, like Mary Ann Masarech and her daughters, make an effort to eat meat from humane and organic farms. Others, like the Farb family, eat no animal products at all. In this chapter we focus solely on the the impact these diets have on animals. What does ethics require of us with regard to eating animals and animal products? In this chapter, the ethics of what we eat become more philosophically complex.

Let's start with factory farming. We have seen how it inflicts prolonged suffering on sows who spend most of their lives in crates that are too narrow for them to turn around in; on caged hens; on chickens kept in unnaturally large flocks, bred to grow too fast, and transported and killed in appalling conditions; on dairy cows who are regularly made pregnant and separated from their calves; and on beef cattle kept in bare dirt feedlots. Though we like and respect Jake and Lee and take into account the time and economic pressures on families with children, we think that buying factory-farm products is not the right thing to do.

You don't have to be a vegetarian to reach this conclusion. Hugh Fearnley-Whittingstall is the author of *The River Cottage Meat Book*—a large, glossy book devoted to the cooking and eating of meat. Yet he writes: "The vast majority of our food animals are now raised under methods that are systematically abusive. For them, discomfort is the norm, pain is routine, growth is abnormal, and diet is unnatural. Disease is widespread and stress is almost constant."[1] Fearnley-Whittingstall lives in England, where laws protecting animals are much stricter than in the United States. American-style crates for sows or veal calves are illegal in Britain, and caged hens have at least 50 percent more space than many American hens are granted. Even so, he considers these conditions abusive to animals. Michael Pollan, another meat-eater, says that factory farms are designed on the principle that "animals are machines incapable of feeling pain" and that to support them requires "a willingness to avert your eyes" from the reality that animals can feel.[2]

Roger Scruton, a critic of animal rights and a vigorous defender of the traditional English sport of foxhunting in the years before parliament banned it, lives on a farm in Wiltshire, where he raises animals for his own table. His attitude to animal rights is perhaps best illustrated by the following incident, as reported by Sholto Byrnes, who visited him at his farm for an interview in *The Independent*:

> After a drink, we move through to begin lunch, components of which have been produced on the Scruton farm. "That's Singer," declares Roger, pointing at a plate of leftover sausages. Singer the pig, mischievously named after Peter Singer, the philosopher and animal-rights theorist, has been "ensausaged" personally by his former owner.[3]

Nevertheless, Scruton flatly rejects factory farming. "A true morality of animal welfare," he writes, "ought to begin from the premise that this way of treating animals is wrong."[4]

In America, those opposed to factory farming include Matthew Scully, a former speech writer in George W. Bush's White House and the author of *Dominion: The Power of Man, the Suffering of Animals, and the Call to Mercy*. Although "animal rights" tend to be associated with those on the left, Scully makes a case for many of the same goals using arguments congenial to the Christian right. In Scully's view, even though God has given us "dominion" over the animals, we should exercise that dominion with mercy—and factory farming fails to do so. Scully's writings have found support from other conservatives, like Pat Buchanan, editor of *The American Conservative,* which gave cover-story prominence to Scully's essay "Fear Factories: The Case for Compassionate Conservatism—for Animals," and George F. Will, who used his *Newsweek* column to recommend Scully's book.[5]

No less a religious authority that Pope Benedict XVI has stated that human "dominion" over animals does not justify factory farming. When head of the Roman Catholic Church's Sacred Congregation for the Doctrine of the Faith, the future pope condemned the "industrial use of creatures, so that geese are fed in such a way as to produce as large a liver as possible, or hens live so packed together that they become just caricatures of birds." This "degrading of living creatures to a commodity" seemed to him "to contradict the relationship of mutuality that comes across in the Bible."[6]

On this issue we agree with Scully, Buchanan, Will, Pollan, Fearnley-Whittingstall, Scruton, and Pope Benedict XVI: no one should be supporting the vast system of animal abuse that today produces most animal products in developed nations.

## UNSOUND DEFENSES OF FACTORY FARMING

What possible arguments can there be in defense of factory farming? We will review some of them and show why they are unconvincing. First, it is sometimes said that we have no duties to animals, because they are incapable of having duties toward us. This has been argued by those who believe that the basis of ethics is some kind of contract, such as "I'll refrain from harming you, if you refrain from harming me."[7] Animals cannot agree to a contract and thus fall outside the sphere of morality. But so, on this view, do babies and those with permanent, severe intellectual disabilities. Do we really have no duties to them either? An even bigger problem for the contract view of ethics is that it cannot ground duties to future generations. We could save ourselves a lot of money and effort by storing radioactive waste from nuclear power plants in containers designed to last no more than, say, 150 years. If we only have duties to those who have duties towards us, why would that be wrong? There is an old joke that goes, "Why should I do anything for posterity? What did posterity ever do for me?" The problem with contract theorists is that they don't get the joke.

Second, when ethical issues are raised about eating meat, many people use what might be called "the Benjamin Franklin defense." Franklin was for many years a vegetarian, until one day, while watching his friends fishing, he noticed that some of the fish they caught had eaten other fish. He then said to himself: "If you eat one another, I don't see why we may not eat you." The thought here may be that if a being treats others in a particular way, then humans are entitled to treat that being in an equivalent way. However, this does not follow as a matter of logic or ethics. Quite rightly, we do not normally take the behavior of animals as a model for how we may treat them. We would not, for example, justify tearing a cat to pieces because we had observed the cat tearing a mouse to pieces. Carnivorous fish don't have a choice about whether to kill other fish or not. They kill as a matter of instinct. Meanwhile, humans can choose to abstain from killing or eating fish and other animals.

Alternatively, the argument could be made that it is part of the natural order that there are predators and prey, and so it cannot be wrong for us to play our part in this order. But this "argument from nature" can justify all kinds of inequities, including the rule of men over women and leaving the weak and the sick to fall by the wayside. Even if the argument were sound, however, it would work only for those of us still living in a hunter-gatherer society, for there is nothing at all "natural" about our current ways of raising animals. As for Franklin's argument about the fish who had eaten other fish, this is a selective use of an argument we would reject in other contexts. Franklin was a sufficiently acute observer of his own nature to recognize how selective he was being, because he admits that he hit upon his justification for eating the fish only after they were in the frying pan and had begun to smell "admirably well."[8]

Third, we have said that the suffering inflicted on animals by factory farming, transportation, and slaughter is unnecessary because—as the Farbs and many other vegan families demonstrate—there are alternatives to meat and other animal products that allow people to be healthy and well-nourished. It might be argued that food from animals is a central part of the standard Western diet and important, if not always central, to what people eat in many other cultures as well. Because animal products are so significant to us, and because we could not buy them as cheaply as we can now without factory farming, factory farming is justifiable despite the suffering it inflicts on animals. But when cultural practices are harmful, they should not be allowed to go unchallenged. Slavery was once part of the culture of the American South. Biases against women and against people of other races have been, and in some places still are, culturally significant. If a widespread cultural practice is wrong, we should try to change it.

It's true that the alternatives to factory farming we've examined, whether Cyd Szymanski's eggs or Niman Ranch pork, are more expensive. Let's grant, too, that switching to a totally vegan diet is something that many people would find difficult, at least at first. But these assumptions are still insufficient to justify factory farming. The choice is not between business as usual and a vegan world. Without factory farming, families with limited means would be able to afford fewer animal products, but they would not have to stop buying them entirely. Nutritionists agree that most people in developed countries eat far more animal products than they need, and more than is good for their health. Spending the same amount of money and buying fewer animal products would therefore be a good thing, especially if those animal products came from animals free to walk around outside, which would make the meat less fatty, and if the reduced consumption in animal products were offset by increased consumption of fruit and vegetables. That is the recommendation of Hugh Fearnley-Whittingstall, and few people are more devoted to food than he is.

For perhaps a billion of the world's poorest people, hunger and malnutrition are still a problem. But factory farming isn't going to solve that problem, for in developing countries the industry caters to the growing urban middle class, not the poor, who cannot afford to buy its products. In developing countries, factory farming products are chosen for their taste and status, not for the consumer's good health. The world's largest and most comprehensive study of diet and disease has shown that in rural China, good health and normal growth are achieved on a diet that includes only one-tenth as much animal-based food as Americans eat. Increases in the consumption of animal products above that very low base are correlated with an increase in the "diseases of affluence": heart disease, obesity, diabetes, and cancer.[9]

The great suffering inflicted on animals by factory farming is not outweighed by a possible loss in gastronomic satisfaction caused by the elimination of meat from animals raised on factory farms from the diet. The harder question is whether we should be vegan or at least vegetarian? To answer that question, we need to go beyond the rejection of unjustified suffering and ask whether it is wrong to kill animals—without suffering—for our food. We need to ask what moral status animals have, and what ethical standards should govern our treatment of them.

## ETHICS AND ANIMALS

The prevailing Western ethic assumes that human interests must always prevail over the comparable interests of members of other species. Since the rise of the modern animal movement in the 1970s, however, this ethic has been on the defensive. The argument is that, despite obvious differences between human and nonhuman animals, we share a capacity to suffer, and this means that they, like us, have interests. If we ignore or discount their interests simply on the grounds that they are not members of our species, the logic of our position is similar to that of the most blatant racists or sexists—those who think that to be white, or male, is to be inherently superior in moral status, irrespective of other characteristics or qualities.

The usual reply to this parallel between speciesism and racism or sexism is to acknowledge that it is a mistake to think that whites are superior to other races, or that males are superior to women, but then to argue that humans really are superior to nonhuman animals in their capacity to reason and the extent of their self-awareness, while claiming that these are morally relevant characteristics. However, some humans—infants, and those with severe intellectual disabilities—have less ability to reason and less self-awareness than some nonhuman animals. So we cannot justifiably use these criteria to draw a distinction between all humans on the one hand and all nonhuman animals on the other.

In the 18th century, Jonathan Swift, the author of *Gulliver's Travels,* made a "modest proposal" to deal with the "surplus" of the children of impoverished women in Ireland. "I have been assured," he wrote, "that a young healthy child well nursed is at a year old, a most delicious, nourishing, and wholesome food, whether stewed, roasted, baked, or boiled."[10] The proposal was, of course, a satire on British policy towards the Irish. But if we find this proposal shocking, our reaction shows that we do not really believe that the absence of an advanced ability to reason is sufficient to justify turning a sentient being into a piece of meat. Nor is it the potential of infants to develop these abilities that marks the crucial moral distinction, because we would be equally shocked by anyone who proposed the same treatment for humans born with serious and irreversible intellectual disabilities. But if, within our own species, we don't regard differences in intelligence, reasoning ability, or self-awareness as grounds for permitting us to exploit the being with lower capacities for our own ends, how can we point to the same characteristics to justify exploiting members of other species? Our willingness to exploit nonhuman animals is not something that is based on sound moral distinctions. It is a sign of "speciesism," a prejudice that survives because it is convenient for the dominant group, in this case not whites or males, but humans.

If we wish to maintain the view that no conscious human beings, including those with profound, permanent intellectual disabilities, can be used in ways harmful to them solely as a means to another's end, then we are going to have to extend the boundaries of this principle beyond our own species to other animals who are conscious and able to be harmed.[11] Otherwise we are drawing a moral circle around our own species, even when the members of our own species protected by that moral boundary are not superior in any morally relevant characteristics to many nonhuman animals who fall outside the moral circle. If we fail to expand this circle, we will be unable to defend ourselves against racists and sexists who want to draw the boundaries more closely around themselves.

## EQUAL CONSIDERATION FOR ANIMALS?

Those who defend our present treatment of animals often say that the animal rights movement would have us give animals the same rights as humans. This is obviously absurd—animals can't have equal rights to an education, to vote, or to exercise free speech. The kind of parity that most animal advocates want to extend to animals is not equal rights, but equal consideration of comparable interests. If an animal feels pain, the pain matters as much as it does when a human feels pain. Granted, the mental capacities of different beings will affect how they experience pain, how they remember it, and whether they anticipate further pain—and these differences can be important. But the pain felt by a baby is a bad thing, even if the baby is no more self-aware than, say, a pig, and has no greater capacities for memory or anticipation. Pain can be a useful warning of danger, so it is sometimes valuable, all things considered. But taken in themselves, unless there is some compensating benefit, we should consider similar experiences of pain to be equally undesirable, whatever the species of the being who feels the pain.

We have now progressed in our argument beyond the avoidance of "unnecessary" suffering to the principle of equal consideration of interests, which tells us to give the same weight to the interests of nonhuman animals as we give to the similar interests of human beings. Let's see whether this principle can help us to decide whether eating meat is unethical.

## EATING MEAT: THE BEST DEFENSE

The most thoughtful defenses of eating meat come from those writers who are strongest in their condemnation of factory farming: Michael Pollan, Hugh Fearnsley-Whittingstall, and Roger Scruton. Pollan's *The New York Times Sunday Magazine* essay "An Animal's Place," begins with the line: "The first time I opened Peter Singer's *Animal Liberation,* I was dining alone at the Palm, trying to enjoy a ribeye steak cooked medium-rare." From there he goes on to describe factory farming and acknowledge that we cannot justify eating the food that this system produces. Pollan then juxtaposes his grim account of modern industral agriculture with a lyrical portrayal of Polyface Farm, spread over 550 acres of grass and forest in Virginia's Shenandoah Valley. Here, Pollan tells us, "Joel Salatin and his family raise six different food animals—cattle, pigs, chickens, rabbits, turkeys and sheep—in an intricate dance of symbiosis designed to allow each species, in Salatin's words, 'to fully express its physiological distinctiveness.'" We learn about Salatin's rotation method: first cows graze on the pasture, then laying hens feast on the grubs attracted by the cowpats, then sheep come and eat the weeds that the cows and hens don't like. There are pigs, too, rooting around in compost in a barn.

If we can recognize animal suffering in a factory farm, Pollan says, "animal happiness is unmistakeable too, and here I was seeing it in abundance." That happiness ends, of course, when the animals are killed, but for the rabbits and chickens, at least, that death is not preceded by the terrifying experience of being trucked off to a slaughterhouse. Salatin slaughters them on the farm. (He would like to slaughter the cattle, pigs, and sheep on the premises, too, but the U.S. Department of Agriculture will not let him.) Salatin's killing is done on Saturday mornings, and anyone is welcome to come along and watch. This leads Pollan to comment that if the walls of both factory farms and slaughterhouses were made of glass, industrial agriculture might be redeemed. Some people would become vegetarians, but others, forced to raise and kill animals in a place where they can be watched, would do it with more consideration for the animal, as well as for the eater. We would have "poultry farms where chickens still go outside" and "hog farms where pigs live as they did 50 years ago—in contact with the sun, the earth and the gaze of a farmer."

In the light of his experience at Polyface Farm, Pollan tells us that to see the domestication of animals as "a form of enslavement or even exploitation" is a mistake. It is, instead, "an instance of mutualism between species" and an evolutionary, not a political, development. Here Pollan may have been influenced by Stephen Budiansky's book *The Covenant of the Wild.*[12] Budiansky's argument is that domestication occurred when some species of animals began to hang around human settlements in order to eat waste or leftover food. Since the animals were edible—or perhaps gave milk and eggs that could be eaten—our ancestors encouraged them to stay around by providing food for them and protecting them from predators. The result has been the evolution of animal breeds that do well, in terms of species survival, by being domesticated. There would be far fewer chickens, pigs, and cattle in the world today if their ancestors had remained wild.

The entire story of domestication is speculative, but one thing is clear: Pollan describes it in a way that cannot be correct and uses it to suggest an ethical justification for our use of animals that it cannot support. He writes that "domestication happened when a small handful of especially opportunistic species discovered through Darwinian trial and error that they were more likely to survive and prosper in an alliance with humans than on their own." No mistake is more common, in accounts of evolutionary processes, than attributing purposiveness either to the process of evolution itself or to entities like genes or species, which are not capable of forming purposes at all. Species do not "discover" anything, through trial and error or in any other way. Individual animals survive and leave offspring, and others, with slightly different characteristics, do not. In this case, on Pollan's account, some animals were attracted to human settlements and were themselves sufficiently attractive to the humans to receive food and protection, while other animals were either not attracted to the human settlements, or were not attractive to the humans. More of the offspring of those animals that were attracted and attractive survived and reproduced than was the case with those animals that were not attracted or not attractive.

Pollan then notes that "Cows, pigs, dogs, cats and chickens have thrived, while their wild ancestors have languished" and that there are now only 10,000 wolves in North America, but 50 million dogs. From this he draws the conclusion that "From the animals' point of view, the bargain with humanity has been a great success, at least until our own time." But just as species are not capable of discovering anything, neither are they capable of making a bargain. Whether individual animals are capable of making a bargain is a separate question, but Pollan is surely not asserting that any individual animal ever consciously made a bargain with humans, to, for example, trade her eggs or milk, or even his or her flesh, for a year or two's food and protection from predators.

Talk of bargains between humans and animals cannot justify anything about how we treat animals today. There is, however, a better point that can be disentangled from Pollan's account of domestication. We can take Pollan to be arguing that since domestic animals have evolved to be what they now are through their symbiotic relationship with humans, their "characteristic form of life"—a phrase Pollan borrows from Aristotle—is one lived in domestication with humans, and that means—for chickens, pigs, cows, and sheep—a life on a farm or ranch. This is their nature, and the Good Life for them is one in which they can live, in accordance with their nature, on the Good Farm, until they are killed and eaten. The killing and eating is unavoidable, for without it neither farms, nor the animals on them, would exist at all.

Fearnley-Whittingstall's defense of meat-eating in *The River Cottage Meat Book* is in some respects strikingly similar to that of Pollan, but it reaches this last point more directly. Fearnley-Whittingstall refers to Budiansky's *Covenant of the Wild* when explaining how "consensual domestication" came about—but he is careful to note that this kind of cooperation between species has nothing to do with individual consent and does not carry the moral implications of individual consent. His point is rather that the nature of farm animals has been shaped by their relationship with us, and they "can be healthy, contented, and even, at least in a sense that suits their species, fulfilled—for the duration of their short lives." Then he adds: "And I believe that these short, domesticated lives are, on balance, better than no lives at all." This gives us moral authority for eating them, but only if we buy from farmers who "embrace the notion of a contract with their meat animals" and "do all they can to uphold it, honourably, morally, and responsibly." *The River Cottage Meat Book* instructs its readers on how to find meat produced by the minority of farmers who do this.[13]

## QUESTIONS ABOUT THE BEST DEFENSE

Pollan's and Fearnley-Whittingstall's defenses of meat-eating are essentially variants on one that is familiar to philosophers who have studied earlier debates about meat-eating. The argument occurs, for instance, in *Social Rights and Duties,* a collection of essays and lectures published in 1896 by the British essayist—and father of the novelist Virgina Woolf—Leslie Stephen. Stephen writes: "Of all the arguments for Vegetarianism none is so weak as the argument from humanity. The pig has a stronger interest than anyone in the demand for bacon. If all the world were Jewish, there would be no pigs at all." Henry Salt, an early advocate of animal rights, thought there was a philosophical fallacy at the core of Stephen's argument: "A person who is already in existence," Salt writes, "may feel that he would rather have lived than not, but he must first have the *terra firma* of existence to argue from; the moment he begins to argue as if from the abyss of the non-existent, he talks nonsense, by predicating good or evil, happiness or unhappiness, of that of which we can predicate nothing."[14]

Salt has drawn our attention to a deep issue that the argument raises. We don't normally think of bringing people into existence as a way of benefiting them. When couples are uncertain about whether or not to have children, they tend to think of their own interests, or perhaps the interests of other existing people, rather than of the benefit they may be conferring on their future children by bringing them into existence—assuming that these children will come into existence in circumstances that make it likely that they will have good lives. But our ordinary way of thinking about such questions might be mistaken. Ask yourself if it would be wrong to bring a child into existence, knowing that the child suffered from a genetic defect that would make her life both brief and utterly miserable for every moment of her existence? Most people will answer "yes." Now consider bringing into existence a being who will lead a thoroughly satisfying life. Is that a good thing to do, other things being equal? If you answer this in the negative, you need to explain why it is wrong to bring a miserable being into existence, but not good to bring a happy or fulfilled being into existence. Sound explanations for this are extraordinarily difficult to find.[15]

We will not attempt to resolve these challenging philosophical questions here. Instead, we'll accept that, as long as a pig has a good life and a quick death, it is a good thing (or at least not a bad thing) for the pig that he or she exists. The argument, then, is that eating meat from farms that give pigs good lives cannot be bad for the pigs, since if no one ate meat, these pigs would not exist. To eat them, however, we have to kill them first, so killing them must be justifiable.

Pollan seems to feel some discomfort about his own argument, because he acknowledges that he has been using what is essentially a utilitarian argument for meat-eating and then recalls that "utilitarians can also justify killing retarded orphans. Killing just isn't the problem for them that it is for other people, including me." So he goes back to Joel Salatin and asks him how he can bring himself to kill a chicken. Salatin replies: "People have a soul. Animals don't. It's a bedrock belief of mine. Unlike us, animals are not created in God's image, so when they die, they just die." As Salatin's answer reminds us, religions often reflects the speciesism of the human beings who developed them. Pollan doesn't comment on Salatin's answer. If he has objections to killing that go beyond utilitarian arguments, he owes us an account of why these objections do not apply to animals.

Fearnley-Whittingstall has noticed that most meat-eaters are protected from thinking about the fact that animals are killed in order to produce meat. He thinks this is wrong, and so he includes in his book a double-page series of color photographs that begins with him taking two of his beef cattle to slaughter, and then shows them being killed, bled out, skinned, disembowelled, and sawn in half. He reports that he watched the process itself and found it "somewhat shocking," although he says that the process "does not seem to me to cause much suffering" and did not make him feel "angry, or sick, or guilty, or ashamed." It compares well, he argues, with almost any other form of death for either a wild or a farmed animal. But Fearnley-Whittingstall doesn't consider that his cattle, like all the animals we eat, died while still very young. They might have lived several more years before meeting one of these other forms of death, years in which they matured, experienced sexual intercourse, and, if they were females, cared for their children. We humans, after all, are prepared to pass up many rapid and humane forms of death in order to live a few more years, even if we are then likely to die of a disease that causes us to suffer before we die.

Scruton's background in philosophy leads him to put his defense of killing animals for food on a more philosophical basis than Pollan or Fearnley-Whittingstall. He writes: "Human beings are conscious of their lives as their own; they have ambitions, hopes, and aspirations." To be "cut short" before one's time is tragic, because "human beings are fulfilled by their achievements and not merely by their comforts." In contrast, animals like cattle do not look forward to future achievements, nor do they seek to achieve anything that will make their lives more fulfilling.[16] Scruton may be right about cattle, but his argument implies that it would be permissible to kill humans who, because of profound intellectual disabilities, are not conscious of their lives as their own and do not look forward to future achievements. Those who find this conclusion too shocking to accept cannot defend the killing of animals for meat on the grounds that animals lack the higher mental abilities that make it wrong to kill normal humans.

## DRAWING CONCLUSIONS

Suppose, though, that some people do accept this disturbing conclusion and eat only humanely raised animals. Does that allow them an impregnable defense of their diet? Not quite. If there were no demand for bacon, nor for any other animal products, farms that now raise animals would convert to growing crops or else go out of business, and humans would replace animal protein with plant protein. Since, as we have seen, we can produce a specified amount of both protein and calories from a smaller area of land when we grow plant foods rather than animal foods, this change would release significant areas of land from agriculture, or would render unnecessary the appropriation of more land for agriculture. If that land were allowed to return to forest, or in the case of existing wild habitat allowed to remain undisturbed, the total number of animals leading lives unconfined by factory farming would increase—for birds and animals are much more abundant in forests than on either cropland or pasture. In North America, for example, there are squirrels, chipmunks, racoons, rabbits, mice, and deer, as well as blackbirds, crows, cardinals, pigeons, sparrows, and starlings—to name just a few. In other countries the species that inhabit forests vary, as do the densities of individual birds and animals, which are highest in tropical forests.

Gaverick Matheny and Kai Chan have attempted to calculate the overall net gain or loss of animal life that will result if people in developed countries should start to switch from their present heavily meat-based diet to one based on plant foods. By calculating the amount of land that could be allowed to return to forest or become some other kind of natural habitat and the number of wild birds and animals who would live on that land, they conclude that even when meat is obtained from grazing cattle living decent lives, the number of animals living free of close confinement will be greater when we obtain protein from plant foods rather than from grazing cattle. The same is true for raising pigs, even if the pigs derive half of their food from waste. In the case of eggs and poultry, with the farming methods like those used at Polyface Farm, the balance may favor continued farming, but this depends on how much grain they need to be fed, in addition to what they can eat on pasture.[17]

Conscientious omnivores might reply that there is no reason to believe that land freed from agricultural use by a switch to a plant-based diet actually would be allowed to revert to wild habitat that could then support the increased number and diversity of animal life. Perhaps it would be bought up for suburban or industrial development. That may be true in some cases in developed countries, especially if the land is near a metropolitan or industrial area. But we should consider the globalized market that now exists for meat. The land no longer needed to produce meat for us may still be used to raise animals whose meat would then be available for export and therefore could slow the rate of forest clearance in, say, Brazil.

There are, of course, exceptions, where animals are raised on land unsuitable for growing crops, and the meat produced is too expensive to be exported. Raising lambs in the Welsh hills, for example, is a traditional form of husbandry that has existed for many centuries and makes use of land that could not otherwise produce food for humans. If the lives of the sheep are, on the whole, good ones, and they would not exist at all if the lambs were not killed and eaten, it can be argued that doing so has benefits, on the whole, for both human and animals.

Pollan also refers to a different argument for eating meat from grazing animals, which he owes to Steve Davis, an animal scientist at Oregon State University. According to Davis, we cannot avoid being responsible for killing animals, even if we are vegan. A tractor plowing a field to plant crops may crush field mice, and moles can be killed when their burrows are destroyed by the plow. Harvesting crops removes the ground cover in which small animals shelter, making it possible for predators to kill them. Applying pesticides can kill birds. Davis then tries to calculate the number of animals killed by growing crops and the number killed by rearing beef cattle on pasture and argues that twice as many animals die per acre when growing crops as in pasture-reared beef production. He then concludes that if we are trying to kill as few animals as possible, we will do better to eat beef—as long as it is fed entirely on grass and not fattened on grain—than to follow a vegan diet.[18] Davis has, however, made a gross error in his calculations: he assumes that an acre of land will feed the same number of people irrespective of whether it is used to raise grass-fed beef or to grow crops. In fact, an acre of land used for crops will feed about ten times as many people as an acre of land used for grass-fed beef. When that difference is fed into the calculations, Davis's argument is turned on its head, and proves that vegans are indirectly responsible for killing only about a fifth as many animals as those who eat grass-fed beef.[19]

Even if it is ethically acceptable to eat animals who have been well-cared for during their lifetimes and then killed without experiencing pain or distress, for those unable to raise their own animals, it is difficult to be sure that the meat you buy comes from such animals. No farm gets more publicity for its exemplary treatment of animals than Polyface Farm. Pollan is not the only one to praise it. The "Style" section of the *New York Times* raved about it and called Joel Salatin, its owner, the "High Priest of the Pasture." Salatin's son has said that his father "has achieved almost godlike status in some circles."[20] But is Polyface really such a good place for animals? Rabbits on the farm are kept in small suspended wire cages. Chickens may be on grass, but instead of being free to roam, they are crowded into mobile wire pens.[21] A review of sustainable poultry systems by the National Sustainable Agriculture Information Service noted that with Salatin's pens "The confined space inside the pens makes bird welfare a concern" and that the crowding "can lead to pecking problems, because the birds lower in the pecking order cannot run away." Out of five sustainable poultry systems investigated, the mobile wire pens were placed last for animal welfare, with a "poor to fair" rating.[22] Herman Beck-Chenoweth, author of *Free Range Poultry Production and Marketing* and a poultry producer himself, calls Salatin's way of raising chickens "a confinement system with a grass floor," adding that although it is "a big improvement over the broiler houses used by companies such as Tyson and Perdue . . . it is a confinement system just the same." [23]

There is also the question of slaughter. The U.S. Federal Meat Inspection Act does not permit Salatin to sell meat from animals that he kills on his farm, so his pigs and cattle are trucked off to conventional slaughterhouses. The crowded transport is likely to be very stressful for them, and it is impossible to know how humanely they are actually slaughtered. Because chickens and rabbits are not covered by the Meat Inspection Act, Salatin can kill them on the farm, sparing them the ordeal of transportation and the strange and sometimes frightening environment of the slaughterhouse. Nevertheless, an account of the killing of chickens at Polyface Farm isn't reassuring:

> Slaughter begins promptly at 8:30 a.m. The goal is to be completely finished by 10:30 a.m. O'Connor, the least skilled of the workers, manhandles the first of 30 crates of birds from a stack on a tractor-drawn trailer outside the pavilion. The birds were taken off of feed and crated about 12 hours earlier so that their craws would be clear for slaughter. He grabs the birds by their feet. Wings flap. Eight white chickens are up-ended in the galvanized metal 'killing cones' at the far end of the processing line. Razor-sharp boning knives flash in the early morning sun. The chickens' throats have been slit. Bright red blood flows down a metal trough and into a large plastic bucket. In a minute or so, the chickens are 'bled out.' They're moved on to the next station in the processing line. And a fresh batch of birds is inserted into the cones.[24]

As this account indicates, birds are crammed into crates with seven other birds—probably including some more aggressive birds they would normally keep away from—and they stay there for 12 hours. Then they are grabbed by the "least skilled of the workers," and passed on, upside down, to other workers who will cut their throats—without any prior stunning. It seems that at Polyface, as elsewhere, it is economics, more than concern for animals, that determines how the animals are treated.

If there are grounds for concern about a farm so often admired, many other supposedly "humane" farms are going to be worse. Not all, of course—we have described visits to some good ones in this book—at least, as far as we could tell from our brief visits. (We were not able to see how any of the animals from these farms were slaughtered.) In practice, as long as animals are commodities, raised for sale on a large scale in a competitive market situation, there will be conflicts between their interests and the economic interests of the producer, and the producer will always be under pressure to cut corners and reduce costs.

Psychological aspects of our choice of diet need to be considered too. Just as farmers who start by raising animals "humanely" may slide into practices more profitable but less humane, so individuals may slide as well. How humane is humane enough to eat? The line between what conscientious omnivores can justify eating and what they cannot justify eating is vague. Since we are all often tempted to take the easy way out, drawing a clear line against eating animal products may be the best way to ensure that one eats ethically—and sticks to it.

The impact we will have on others is even more important. Since factory farming inflicts a vast quantity of unjustifiable suffering on animals, persuading others to boycott it should be a high priority for anyone concerned about animals. In this respect, a broad brushstroke may be better than a more finely-tuned approach. Vegans and vegetarians draw clear lines by refusing to eat all, or some, animal products. Whenever they dine with others, that line is evident, and people are likely to ask them why they are not eating meat. That often leads to conversations that influence others, and so the good that we can do personally by boycotting factory farms can be multiplied by the number of others we influence to do the same. When conscientious omnivores eat meat, however, their dietary choices are less evident. On the plate, ham from a pig who led a happy life looks very much like ham from a factory-farmed pig. Thus the eating habits of the conscientious omnivore are likely to reinforce the common view that animals are things for us to use and unlikely to influence others to reconsider what they eat.

Where does all this leave the diet of conscientious omnivores? Perhaps it's not, all things considered, the best possible diet, but the moral distance between the food choices made by conscientious omnivores and those made by most of the population is so great that it seems more appropriate to praise the conscientious omnivores for how far they have come, rather than to criticize them for not having gone further.

## KILLING YOUR OWN

Farms are not the only source of meat. Fearnley-Whittingstall's *River Cottage Meat Book* includes hints on cooking pheasant, partridge, pigeon, mallard and teal ducks, geese, grouse, woodcock, snipe, rabbit, hare, and venison, with a reference to cooking "the odd squirrel" as well. It seems safe to say that many readers, including many who eat meat, will be repulsed by this list. Although only 4 percent of Americans say they are vegetarian, opinion polls show that at least three times as many are opposed to hunting, even for food.[25] In other countries, with less of a gun-owning culture, opposition to hunting is probably stronger still. Yet when compared with the factory-farmed chickens that most people eat in such vast numbers, the wild birds shot by hunters have a far better life and usually a much quicker death. Unlike factory-farmed pigs, wild boar grow up with their mothers and are able to move around freely. Wild deer never have to endure 36 hours crammed into trucks without rest, food, or water, on the way to the feedlot or to slaughter. Is meat from wild birds or animals really worse, ethically, than meat you buy in a supermarket?

In Fearnley-Whittingstall's opinion, "truly wild animals, dispatched efficiently by a good shot, provide us with meat that is perhaps the least ethically problematic of all."[26] He admits, however, that not all "game" meets this standard. First, a lot of it is not truly wild. In Britain, for instance, pheasants are hatched in incubators, and for the first few weeks of their lives are raised just like factory-farmed chickens, in big sheds. They are then moved to pens in woodland, where they are still fed on grain, before being released for the benefit of shooters who pay large sums for the "sport" of killing them. Fearnley-Whittingstall thinks that this is still a good life for a pheasant and that shooting and eating them is ethically defensible. One might doubt that. Second, even though "most" birds and mammals shot by hunters are "dispatched efficiently," Fearnley-Whittingstall admits that "even the best shots will sometimes miss, and sometimes wound the birds and mammals they are trying to kill. The worst shots will do so frequently." These animals "will die of their injuries hours or even days after being shot. And there is no kidding ourselves that those hours or days will not be spent in some pain."

As with farm animals, Fearnley-Whittingstall argues that part of the ethical justification for hunting is a kind of symbiotic contract between wild animals and us. This time the basis of the contract is that we manage the habitat in which they live. "There is barely a square metre of Britain left—and not much of the whole world, come to that—that is not made the way it is by our interference, or at the very least allowed to remain the way it is by our concerted effort not to interfere." Since human emissions of greenhouse gases have already affected our planet's climate to some extent, it is, strictly speaking, true that there is no part of the world unaffected by human action. But it is hard to see that this kind of interference creates a contract between wild animals and ourselves. And if we refrain from cutting down forests and using the cleared land for grazing cattle, does that mean that we have benefited wild animals in a way that entitles us to kill and eat them? That sounds like a more powerful nation saying to a weaker one: "We could kill you all and take your land, but since we have decided not to do so, we have benefited you and you should therefore work on our plantations." Unlike animals raised on farms, the animals killed by hunters would have existed quite independently of us, and their deaths mean that there are fewer animals enjoying their lives. The desire to cook a pigeon rather than some beans is not enough reason to end the pigeon's enjoyment of his or her life.

A better case for hunting can be made when the animals hunted are causing ecological problems. Rabbits introduced to Australia by European settlers have changed Australia's unique plant and animal life and threaten native species with whom they compete. Australian possums, taken to New Zealand for fur farming, have escaped and multiplied prodigiously in New Zealand's forests, which have evolved in the absence of leaf-eating mammals. If these animals are going to be killed anyway to protect the environment, it is difficult to see any objection to eating meat taken from their bodies. There is a danger that the desire to hunt and eat the animals will make shooting them the preferred means for dealing with the environmental issue, when other less harmful means (like forms of sterilization) could be used. Whether hunting can, in limited circumstances, be justified depends on whether such other means are available, or could be developed.

## DUMPSTER DIVING: THE ULTIMATE ETHICAL CHEAP EATS

Even products that contain no animal ingredients can hurt animals, when land is cleared to grow crops, or when oil companies go into wilderness areas to provide the fuel needed to truck the goods around the country. There are some who take a more radical approach: to minimize their impact on animals and on the environment, they live on what supermarkets throw out. They are known as "freegans," a term that is a deliberate play on "vegans." To see what food it is possible to gather in this way, we accepted an invitation to join a foraging expedition in one of Melbourne's northern suburbs.

It's about 7.30 p.m. on a mild Tuesday evening. We're in a small Toyota station wagon with Tim, Shane, G (Gareth), and Danya. They're all in their 20s, wearing old denim or waterproof jackets, except for G, who is wearing a jacket that might once have been more stylish and formal, but is now so worn that it would have suited Charlie Chaplin in "The Tramp." The comical appearance is reinforced by the fact that G is tall and lanky, and this jacket was made for someone much smaller. We park in the Safeway parking lot, but avoid the customer entrance, heading instead to the delivery ramp. A dumpster bin stands at the side. The lid is chained and locked, but the chains have enough slack to allow you to raise the lids and insert an arm. G and Danya get their arms in and start bringing out loose potatoes, plastic wrapped packages of broccoli, a bunch of asparagus, plastic packs of flat Lebanese bread, and a small can of tuna. The tuna can is dented, the broccoli is looking a little tired, and some of the potatoes have a slight greenish tinge. We collect what we want, throw the rest back, replace the plastic bags and other trash that has come out accidentally, and leave the area at least as tidy as we found it.

We move around the corner to where there is another bin, this time unlocked. We throw the lid open to reveal boxes of strawberries. Tim says strawberries are not worth taking, they usually taste bad. Instead he

## ANIMAL-FREE MEAT?

> "Skum-skimming wasn't hard to learn. You got up at dawn. You gulped a breakfast sliced not long ago from Chicken Little and washed it down with Coffiest. You put on your coveralls and took the cargo net up to your tier. In blazing noon from sunrise to sunset you walked your acres of shallow tanks crusted with algae. If you walked slowly, every thirty seconds or so you spotted a patch at maturity, bursting with yummy carbohydrates. You skimmed the patch with your skimmer and slung it down the well, where it would be baled, or processed into glucose to feed Chicken Little, who would be sliced and packed to feed people from Baffinland to Little America. Every hour you could drink from your canteen and take a salt tablet. Every two hours you could take five minutes. At sunset you turned in your coveralls and went to dinner—more slices from Chicken Little—and then you were on your own."[27]

That is Frederick Pohl and C. M. Kornbluth's fantasy, in *The Space Merchants*, of how we might one day produce our food. "Chicken Little" is a vast lump of meat, hundreds of feet across, growing in a culture that is fed on algae. The idea has attracted more eminent and realistic figures than science fiction writers. In 1932, Winston Churchill wrote: "Fifty years hence, we shall escape the absurdity of growing a whole chicken in order to eat the breast or wing, by growing these parts separately under a suitable medium."[28] Churchill was better at predicting Hitler's aggressive intentions than the future of food production, but it may be only his timing that was astray. Within the animal movement, there are some who hope that, just as the development of the internal combustion engine has eliminated the suffering of millions of horses and oxen previously used to transport people and heavy goods, so eventually the development of *in vitro* meat will eliminate the suffering of billions of animals now used for meat.

picks out some tomatoes and capsicums, two large bottles of orange juice, wholemeal loaves of bread, white rolls, packs of croissants, and maybe thirty packets of flat bread. Shane comes up with a long piece of fish. "Ah, Blue Grenadier!" he says, but he's laughing, because it smells really bad, and he throws it back. "There's some really skank stuff down

We already have vegetarian burgers, sausages, bacon bits, and many other meat-like products. In China, when the spread of Buddhism led to abstention from meat during religious festivals, the Emperor's chefs devised ways of making gluten and tofu resemble various forms of meat and seafood, so that the Emperor could continue to dine on the classic dishes of Chinese cuisine. Today, this tradition is still practiced in Chinese vegetarian restaurants. Some vegetarians object to "fake meat" because it may leave the impression that meat dishes set the culinary benchmark, and some meat-eaters don't like it because the taste and texture is not the same as that of meat. *In vitro* meat would not be fake meat, it would really be meat, with an identical taste and texture. In theory, growing meat in culture should be more efficient than producing entire animals, since as Churchill suggested, we should be able to grow pure boneless steak or chicken breast, without producing inedible bone, unhealthy fat, and undesirable internal organs. Cultured meat should also have less impact on the environment than factory farms because it would not produce any manure.

Scientists can already produce small amounts of muscle tissue in a laboratory. In 2001, a scientist from the University of Amsterdam, together with two Dutch businessmen, took out a patent on a process for producing meat by bathing muscle cells in a nutrient solution and inducing them to divide. Several scientists in the United States as well as the Netherlands are working on producing edible meat, but so far without success. It may eventually prove possible; the real question is, whether it will be possible for such meat to compete economically with meat from living animals.

One scientist has estimated that the current cost of producing muscle tissue in a laboratory equates to $5 million per kilogram! But only 50 years ago, the exorbitant cost of building a computer meant that few people imagined that they would ever become affordable to ordinary families. If, one day, cultured meat becomes an efficient way of producing food, we see no ethical objection to it. Granted, the original cells will have come from an animal, but since the cells can continue to divide indefinitely, that one animal could, in theory, produce enough cells to supply the entire world with meat. No animal will suffer in order to provide you with your meal.[29]

here," Danya warns, "watch out for the orange plastic bags, they're full of bad meat." We pick up the pile of food we have collected, but put back most of the bread, keeping just a few of the flat breads and one pack of croissants. "We don't need that much, and there might be others coming after us," Tim explains.

Shane and G have gone somewhere else and return with cartons full of small bottles of orange juice. But they sample one, and it's fizzy. They try another, same thing. It goes back in the dumpster. We head off to another group of supermarkets. Danya, who is sitting next to Shane, complains about the bad fish smell that still lingers on Shane's hands.

The next bin we visit is standing by the loading dock and isn't locked, so this time we climb up onto the dock and investigate the contents from above. Danya is delighted to find several cakes, still in clear plastic display boxes. It's her 21st birthday today, and she claims one as a birthday cake. She also finds a tray of chicken breasts and one of chicken drumsticks. "Are they cold?" Tim asks. Yes, they still feel cold. That means they haven't been out of refrigeration long, and that makes them acceptable. Two dozen eggs, still in their cartons, are another find worth taking. So too are a bag of sugar, some tins of tomatoes, a large pack of Chinese noodles, and a torn bag of pasta shells. "Does anyone drink Coke?" G asks. He's found a pack of 24 cans. Shane says yes, he'll have them. The carton is ripped, but the cans are intact, so he starts loading them into another cardboard box. G pulls out a chocolate cake covered in cream, removes the plastic packaging, takes a large bite, and pronounces it good. There are large packs of toilet paper. "That's good, we always need them," says Shane. There is even an electric toothbrush, still in its package.

While we are going through the bin, the roller door behind us starts moving, and an employee who looks about 16 comes out with a wheelie bin to empty into the dumpster. He doesn't look particularly surprised to see us there, but he says, "If security comes around you'll be in trouble." Tim nods assent and offers assistance in unloading the contents of the wheelie bin into the dumpster. The exchange is polite and friendly. We never see any security people, and this turns out to be the only encounter with anyone from one of the stores this evening. That's fairly typical. If they are asked to leave, they say, they just go.

We move on to another supermarket up the road. The bins here are chained down again, but the gap is wide enough for G to spot some coffee he wants. It's too far down for even his long arm to reach, so for the first and only time tonight we see some real "dumpster diving" as G gets his upper body right into the bin, only his legs sticking out the top. The booty is eight 250-gram vacuum-sealed packets of an imported Italian Arabica coffee, just a couple of days past the expiration date.

By now the back of the car is getting full and we are hungry, so we head home to cook. Home is an office-warehouse building that the group has been squatting in for about six months—apparently there is a legal dispute about who owns it, and the property had been vacant for years before they moved in. They've furnished it almost entirely with discarded items and had electricity and gas put back on, so it's comfortable and very spacious. Tonight Tim does most of the cooking, with some advice and assistance from others. He chops up the asparagus, zucchini, broccoli, and fresh tomatoes and opens two tins of tomatoes as well. That all goes into a pot and gets cooked up. Meanwhile the pasta is boiling away, and when everything is ready we serve ourselves some pasta and add the sauce. If this were a restaurant they'd probably call it "Pasta Primavera." We wash it down with sparkling organic apple juice from New Zealand, in individual bottles. We've had better meals in restaurants, but we've had worse too.

Although some of the items we got were past their use-by date or had damaged packaging, with others there was no obvious reason why they had been thrown out. The expiration date on the eggs was still two weeks in the future and none were broken. The cans of Coke and the Chinese noodles weren't damaged or about to go bad. The toilet paper and electric toothbrush would have lasted indefinitely. "You find stuff and can't figure out why it has been thrown away," Tim says, "We got cartons of organic breakfast cereal and the use-by date was two months ahead." "And what about this organic apple juice?" we ask, holding up the bottle we've just enjoyed drinking. "That had a use-by date about a year ago," Tim says, and everyone laughs at our evident discomfort. "But don't worry, it's perfectly fine." And indeed it was. We experienced no after-effects, from that or any of the other ingredients in the meal. Nor have any of the others ever had any stomach problems from a dumpster meal.

After we've eaten, Danya goes out with a friend to celebrate her birthday, and the rest of us start talking about lifestyles and "dumpstering." G says he got started about two years ago when he was reading Georges Bataille, the French writer and thinker who died in 1962. In contrast to conventional economists, who start from the problem of scarcity and how best to overcome it, Bataille analyzed the prevailing social and economic order by seeing what it does with its excess. So the next time G passed a supermarket's dumpster, he looked in. "There were about a hundred bananas in there," he said. That got him really excited, and he has been dumpstering ever since. Now he gets all his meals from dumpstering, living from the excess of corporate capitalism. Some days are better than others, he says, but you can always find a meal. G is studying at university. In Australia, students without enough money to live on are eligible for government financial support, but G feels no need for it and hasn't applied. "I can live without money."

Tim takes a slightly different view. He earns some money, but goes dumpstering in order to save for things he can't get free. "It's a question of priorities. Beyond wanting to save money, it's about how you want to spend it. Whether you want to be just a mindless consumer, or whether you want to put your money into useful stuff, and save money for things that are tools, like keeping a car on the road that I need, buying laptops, and digital video players. It means that you are able to have access to resources that we couldn't otherwise afford."

Shane has been dumpstering for about five years. Dumpstering, he says, is empowering. "Think of the single mother who has to scrape together enough money to be able to buy a tin of baked beans and some white bread for herself and her kid. If she had the confidence to go around the back and walk up to the bin, she could get much better food for nothing. But she can't transcend the cultural shame. For us, it's culturally acceptable to do it, and we have the skills and the confidence as well. So although none of us has a high-paying job, we live a very comfortable lifestyle, much better than we could afford on what we earn if we had to pay for everything we use."

Tim says it's important to think about dumpstering in its political and economic context. "We have to get away from the simplistic idea that 'you're eating out of a bin, therefore you're a dero,'" he says. "That's going to prevent the single mother from getting food out of the bin. We have the political analysis that enables us to rise above that way of thinking." Shane agrees: "What's better about dumpstering is that you're not buying into that whole process of consumption. Even buying organic food involves being part of the consumer economy. Dumpstering really does break the consumer chain."

"But the people who buy organic could say that they're changing the system," we interject. "By giving money to organic farmers, they're encouraging farmers to grow more organics. What you're doing isn't changing the system, you're just living off a glitch in the system." That provokes Shane and Tim to a critique of the extent to which organic farming has become part of the system of agriculture and marketing, rather than a real alternative to it. Shane acknowledges that some forms of community supported farming and organic cooperatives could be a real alternative, but in his view, the "local yuppie organic store" certainly isn't. Dumpstering is much more radical. "It's an act of withdrawal—a withdrawal from the whole process of industrial food production and marketing."

G comes back into the discussion, saying that dumpstering has "an ethical dimension . . . We're saving food that would otherwise totally go to waste—perfectly good food. We're recycling it." Tim adds: "It's got to be the lowest-impact form of food consumption." Then he goes on to say that because you don't need much money, you can spend your time doing something socially useful, rather than getting a meaningless job to earn money to buy food. To judge by the leaflets and notices stuck on boards in the house, people living there are spending time on campaigns for indigenous Australians, against duck shooting, for environmental protection, and against the war in Iraq.

Apart from all that, this way of getting food just seems to be a lot of fun. 'It's a daily victory against the system," Tim says. "Every day you come home and think 'I've won.' It may be only a small victory, but I've won." Shane talks about the "rush" of finding great stuff, and G mentions the communal aspect of dumpstered meals: "There's a really good alternative economy in terms of the way you can share and distribute your food as well. Every meal you can share with a couple of people, and there's never any hassle or concern about where the food is coming from. You know it is from this resource that is kind of . . . unending. It's a permanent gift." G also relishes the challenge of getting a few things and working out what you're going to cook with them. That sparks a lot of reminiscing about the good things they've found, and, amidst laughter, they discuss the great and the not-so-good meals they've made entirely from dumpstered food.

Our evening of dumpstering in Melbourne could have been replicated in any large city in the U.S., Canada, or much of Europe. Nobody knows how many people do it, but at the time of writing, www.Meetup.com listed 1,888 people interested in dumpstering, and the New York City group alone had 199 members.[30] We had imagined that dumpstering would retrieve only old or blemished food and were astonished by the non-perishable items in perfect condition we found in dumpster bins; later we discovered that our gleanings were typical of what is thrown out in many countries. A New York dumpster diver recounts finding dumpsters full of expensive packages of gourmet nuts and dried fruit, luxury chocolates, three or four 50-pound bags of bagels regularly thrown out by a single deli, and large quantities of non-perishable food like rice pilaf mixes and instant soups.

Some of this waste is easily explained. Bakeries, donut stores, delis, and salad bars often advertise that they bake fresh, or get freshly made food every day, and they also like to keep their racks and salad bowls full, so that customers don't get the impression that they are buying the dregs after other customers have picked them over. This combination ensures that at the end of the day a lot of perfectly good food gets thrown out. A small fraction of it may be donated to food banks or shelters for homeless people, but most of it is simply put in the bin, probably because the stores are worried about undermining their own sales—if the word gets around that you can get something for free at 10 p.m., fewer people will buy it at 8 p.m. But the reasons for trashing non-perishable goods are more mysterious. On some products, stores get lower prices for ordering large quantities, so it can be cheaper for them to order more and put what they don't sell in the trash than to buy only what they can sell. Perhaps more importantly, shelf space is a limited resource, and stores regularly clear out shelves for new deliveries. The store may have a long-term contract with a supplier to provide a specified quantity of a product each week. If an item has not sold as well as expected, the old stock will be dumped, even if it is not out of date, to make way for the new stock.

Many dumpster divers began as vegans, but became convinced that boycotting animal products is not radical enough. An anonymous vegan has said that being a freegan means that "you are boycotting EVERYTHING! . . . That should help you get to sleep at night."[31] While freegans are more radical than vegans in refusing to purchase any kind of food at all, they are also more flexible, in that they see no ethical objection to eating animal products that have been thrown out. They want to avoid giving their money to those who exploit animals. Once a product has been dumped, whether it gets eaten or turned into landfill can make no difference to the producer. Some freegans still don't like the idea of dining on a corpse, and—although they are prepared to eat food from dumpsters—they know about fecal contamination on meat and see health risks in eating anything that has passed through a slaughterhouse. But their reasoning is impeccably consequentialist: if you oppose the abuse of animals, but enjoy eating meat, cheese, or eggs—get it from a dumpster.

Freeganism is not only about free food. Behind it lies a view about how to live one's life, one that rejects the priorities set by the consumer society and the lifestyle that results from accepting those priorities. Because most people see their status as linked to wealth and what they can buy, they are locked into working, often in unsatisfying jobs, to earn the money they need to enhance their status. Freegans reject that idea of status and do not even need to earn money to satisfy their basic needs. They point out that we all have far more consumer goods than people did in the 1950s—when most people had smaller homes, and no one had DVDs, microwave ovens, cell phones, or personal computers—and yet surveys show that we are no happier now than we were then. Freegans see happiness as something that comes from doing things, rather than having things. If they work at all, it will be because they see the work they are doing as worthwhile in itself. To a far greater extent than people who pay for everything they consume, freegans' time is, as Tim said, their own, to enjoy or to use for working for what they believe in. They are thus doubly free—free from subordination to the consumer ethos and free from the need to work to satisfy their needs. They think that an alternative, less exploitative economic system is possible, but they are under no illusion that taking food from dumpsters will in itself bring that system about. Instead they see dumpster diving both as a way of detaching themselves from the present system and, at the same time, as part of a broader life of resistance to that system.[32]

Dumpster diving may not be an option many consumers are likely to explore, but there's still a lesson to draw. Many of agriculture's ill effects on laborers, animals, and the environment could be reduced if we ate what would otherwise be wasted. According to Dr Timothy Jones, an archaeologist at the University of Arizona who led a U.S. government-funded study of food waste, more than 40 percent of the food grown in the United States is lost or thrown away—that's about $100 billion of wasted food a year. At least half of this food, Jones says, could have been safely consumed. Waste could also be reduced by having better storage facilities. Some of the waste is completely pointless and reflects nothing more than a casual disregard for what went into producing the food, from the suffering of the animals, to the labor of the workers, to the natural resources consumed and the pollution generated. Jones examined what stores, restaurants, and individuals throw out and found that 14 percent of household garbage was perfectly good food that was in its original packaging and not out of date. About a third of this edible food was dry-packaged goods, and canned goods that keep for a long time made up another 19 percent. Jones speculated that discounts for bulk purchases lead people to buy more food than they want to keep, but he admits to some bafflement, remarking: "I just don't understand this."[33] As consumers, we have direct control over our own waste. We'd do well to follow the advice our mothers gave us: eat your leftovers.

# 18

# WHAT SHOULD WE EAT?

Before we reach specific conclusions about how we should eat, we will outline five ethical principles that we think most people will share. These principles do not encompass everything that is morally relevant, but they can help us to decide all but the most contentious ethical issues.

**1. Transparency: We have a right to know how our food is produced.**
If slaughterhouses had glass walls, it's often said, we'd all be vegetarian. That's probably not quite true—some people can get used to almost anything. But transparency is increasingly recognized as an important ethical principle and a safeguard against bad practice. Consumers should be able to get accurate and unbiased information about what they are buying and how it was produced.

**2. Fairness: Producing food should not impose costs on others.**
The price of food should reflect the full costs of its production. Then consumers can choose whether they want to pay that price. If no one does, the market will ensure that the item ceases to be produced. Meanwhile, if the method of producing food imposes significant costs on others without their consent—for example, by emitting odors that make it impossible for neighbors to enjoy living in their homes—then the market has not been operating efficiently and the outcome is unfair to those who are disadvantaged. The food will only be cheap because others are paying part of the cost—unwillingly. Any form of food production that is not environmentally sustainable will be unfair in this respect, since it will make future generations worse off.

**3. Humanity: Inflicting significant suffering on animals for minor reasons is wrong.**

Most people, even those opposed to more radical ideas of "animal liberation" or "animal rights," agree that we should try to avoid causing pain or other forms of distress on animals. Kindness and compassion toward all, humans and animals, is clearly better than indifference to the suffering of another sentient being.

**4. Social Responsibility: Workers should have decent wages and working conditions.**

Minimally decent treatment for employees and suppliers precludes child labor, forced labor, and sexual harassment. Workplaces should be safe, and workers should have the right to form associations and engage in collective bargaining, if they so choose. There must be no discrimination on the basis of race, sex, or disabilities irrelevant to the job. Workers should receive a wage sufficient to cover their basic needs and those of dependent children.

**5. Needs: Preserving life and health justifies more than other desires.**

A genuine need for food, to survive and nourish ourselves adequately, overrides less pressing considerations and justifies many things that might otherwise be wrong. In contrast, if we choose a particular food out of habit, or because we like the way it tastes, when we could have nourished ourselves equally well by making a different choice, then that choice has to meet stricter ethical standards.

Drawing on these principles, let's look at some of the food choices, bearing in mind the information provided in earlier chapters.

## FACTORY-FARMED FOOD

In supermarkets and ordinary grocery stores, you should assume that all food—unless specifically labeled otherwise—comes from the mainstream food industry and has not been produced in a manner that is humane, sustainable, or environmentally friendly. Animal products, in particular, will virtually all be from factory farms, unless the package clearly states the contrary. Don't be fooled by terms like "all-natural" or "farm fresh." They are often used to describe factory-farm products.

**Factory-farmed Chicken:** The first food purchase we examined—Jake and Lee's chicken—is one of the worst. Whether the chicken comes from Tyson, Gold Kist, Perdue, or any of the intensive producers, the conditions in which the birds live, are transported, and die should be enough to disqualify factory-farmed chicken from every ethical shopping list. In addition, broiler sheds have polluted water resources and ruined the lives of people into whose area they have moved. Work in chicken slaughterhouses is dirty, dangerous, and low-paid, and attempts to unionize plants may be met with intimidation.

**Factory-Farmed Turkey:** Everything we have said about factory-farmed chicken applies to factory-farmed turkey. Think about it before Thanksgiving. We prefer "Tofurky," a tofu "turkey" made from organic soybeans that has a "skin" that turns deliciously crisp when baked. (See www.tofurky.com for more information.) Those who simply must have a real turkey should at least seek out one who has been organically raised, with access to pasture.

**Eggs from Caged Hens:** Everything we have said about factory-farmed chicken applies to eggs from caged hens. The lives of the closely-confined hens are, if anything, even more miserable than those of broilers. No one needs to eat eggs from caged hens. For those on a limited budget, a healthy option is to eat fewer eggs and buy the more expensive but better-tasting eggs from hens free to move around inside sheds or, preferably, outdoors (as described in Chapter 8). The other choice is to replace all eggs with vegan alternatives—for cooking and baking, there are several vegan egg substitutes, available in natural food stores or online.

**Factory-Farm Veal:** Most people who think about food ethics have already crossed off this item, partly because of the narrow stalls—too narrow for them even to turn around—in which they are still reared in many counries. Keeping veal calves in individual stalls is now banned in Britain, and will be banned in the European Union from 2007. Even so, the calves suffer from early separation from their mothers, and can still be kept entirely indoors, fed on an artificial milk replacer. The food we give them, which could be eaten directly by humans, contains far more calories and protein than the meat we get from them.

**Factory-Farm Pork, Ham, and Bacon:** In most developed countries, sows on pig farms are treated as breeding machines and are so closely confined that they cannot even turn around, let alone walk. Individual sow stalls are now banned in Britain. In the European Union, they will be phased out by 2013, but the regulations allow for the confinement to continue for the first 4 weeks of each pregnancy. Their offspring are permanently indoors, on concrete, without bedding. Intensive pig farms use 6 pounds of grain for every pound of boneless meat they produce, putting stress on the environment. They use large amounts of fuel and often generate considerable pollution problems too. We don't consider this an ethical food choice.

**Factory-Farm Milk, Cheese, and other Dairy Products:** Intensively-reared dairy cows are genetically selected and managed so that for a few years they will give vast quantities of milk. To that end they are regularly made pregnant and separated from their calves soon after birth, which causes distress to both mother and calf. They don't get to walk around on pasture and eat grass. If their bodies collapse and they can no longer walk, they are considered worthless and, as we have seen, may be tied to a tractor and dragged across concrete and into a truck before being killed. Many male calves of dairy cows go to the veal industry. In addition, intensive dairy farms can be serious polluters. So intensively produced dairy products should be avoided. (Unfortunately, most of these problems occur in large-scale organic dairy production as well.)

**Intensively-Produced Beef:** Beef is not a factory-farmed product in the same way as the other animal products we have considered. Calves raised for beef usually live for at least six months by their mothers' side, on pasture. Later, in feedlots, they have more room to move than chickens, laying hens, veal calves, or pigs and they are under less stress than dairy cows. But they are still subjected to extreme hot and cold weather without shade or shelter, as well as hot iron branding, dehorning, and castration, all without anaesthetic. They can, under an 1872 United States law, be transported by rail for 28 hours without water or rest—and even that minimal protection has been rendered obsolete by the widespread use of trucks to transport cattle. The U.S. Department of Agriculture refuses to interpret the "Twenty-Eight Hour Law" as applying to trucks, which carry 95 percent of the farm animals transported, so there are effectively no limits on how long calves can be trucked without water, food, or rest. Cattle find their feedlot diet difficult to digest. It gives them diseases from which they would die without antibiotics—and even the antibiotics would not suffice to keep them alive much beyond the age at which they are slaughtered.

The lives of feedlot beef cattle are probably still better, on the whole, than those of factory-farmed pigs or chickens. The counterweight to that, however, is the inefficiency of cattle as converters of grain to meat. This means that eaters of feedlot beef are responsible for the use of more land, fertilizer, fossil fuels, water, and other resources than those who eat chicken. They may also be responsible, as we saw in Chapter 16, for the degradation of public lands.

**An Overall Verdict on Factory-Farmed Meat, Eggs, and Dairy Products:** As explained in Chapter 16, concentrated animal feeding operations reduce the amount of food available for human consumption. We don't need them. What factory farms do to animals, nearby residents, and the entire planet's environment, they do because people are accustomed to eating these animal products and can't imagine a meal without them, or because they like the way they taste. These are not ethical justifications, given the harm these practices cause. Supporting factory farming by knowingly buying its products is wrong.

There are many alternatives to factory farm products. If price and convenience are important, tofu and vegan soyburgers are similar in cost, for the protein they contain, to factory-farmed chicken, while dishes based on dried beans and lentils provide more protein for less money. Many vegans and vegetarians recommend Boca "chicken style" patties and chicken nuggets. Textured vegetable protein ordered from www.healthy-eating.com is another bargain-priced source of protein.[1] Or you could just eat less protein. As we have seen in Chapter 15, adults who get enough calories will almost always be getting enough protein, without specifically seeking high-protein foods.

Animal products from sustainable producers who show some concern for animal welfare are available in many areas. Start by checking out the health food and natural food stores and food co-ops in your area. If they don't stock what you are seeking, ask them to do so. (And then be sure to go back and buy it!) If you live in a small town or rural area where there are no such stores, look for farmers' markets and buy directly from local farmers who are open about their methods and will allow you to visit their farm.

A growing variety of food is available for delivery to your door, often as part of a Community Supported Agriculture, or CSA, scheme. Got to www.cuco.org.uk for a clickable map that will tell you about farmers in your area who are part of such a scheme. You can also order organic, fair trade, and locally produced food to be delivered to you from www.ethicalfoods.co.uk. These and other useful resources can be found at the back of this book.

In the United States, for example, see www.healthy-eating.com or write to The Mail Order Catalog, 413 Farm Road, P.O. Box 180, Summertown, TN 38483. You can also try the Eat Well Guide, a free, online directory of sustainably raised meat, poultry, dairy, and eggs from farms, stores, restaurants. and online outlets in the U.S. and Canada. By going to www.eatwellguide.org and entering your zip code, you may be able to find local outlets for products from animals who were raised organically, sustainably, and on pasture. A list of other useful resources can be found at the back of this book.

## FISH AND OTHER MARINE ANIMALS

**Farmed Fish:** Fish farming is factory farming in the water, and, like land-based forms of factory farming, it suffers from the general problem of concentrated animal feeding operations—we have to catch or grow the food and bring it to the animals we are feeding. With fish farming, the extent to which this damages the oceans and wastes food resources depends on whether the fish are carnivorous, as salmon are, or herbivorous, like carp. Carp are farmed and eaten extensively in China, where they contribute to meeting the protein needs of rural people with relatively few food choices. In the industrialized countries, however, it is usually the more expensive carnivorous fish that are eaten, and farming them wastes the ocean's resources. Fish farming often causes pollution problems too. Farmed fish may be stressed from the crowding and confinement to which they are subjected. The methods by which they are killed show total indifference to their pain and suffering. For these reasons, farmed fish isn't an ethically acceptable food.

**Wild-Caught Fish:** If you are going to eat fish at all, wild-caught fish is definitely preferable to farmed fish, as long as it comes from sustainable fisheries. Look for the Marine Stewardship Council's "Fish Forever" seal, or check the species you are buying on "The Fish List" (www.thefishlist.org), a Web site maintained by Environmental Defense, Blue Oceans Institute, Monterey Bay Aquarium, and Seafood Choices Alliance. But identifying appropriate species may not be enough, because even high-end stores have been caught selling farmed salmon mislabeled as wild salmon. That means that well-intentioned consumers seeking to make an ecologically good choice could instead unwittingly be buying one of the ecologically worst choices.

Suppose, however, that you buy fish that, like Jake and Lee's pollock, is from a sustainable fishery. The fish live free, so eating them doesn't contribute to prolonged suffering, as it does when you buy factory-farmed animal products. To that extent, eating fish can be a better ethical choice than eating most of the meat, eggs, and cheese sold in supermarkets. Still, the suffering fish experience while dying is a compelling reason for avoiding fish, at least for families able to get enough food, and sufficient protein, without eating fish.

**Invertebrates:** The key ethical issues related to eating invertebrates (such as squid, octopus, crabs, lobster, shrimps, oysters, clams, and mussels) are environmental sustainability and the possibility of causing unjustified suffering. On the environmental question, there are too many species and too many ways of farming or catching them to detail here, and any information we might provide would soon be outdated by changing conditions and new information. The Fish List site mentioned on the previous page is a reliable source of guidance, but to avoid contributing to environmental damage to the oceans and seabed, it is essential to know what you are eating and where it comes from.

Shrimp, now America's number one seafood, is mostly imported and produced in an unsustainable manner. It should therefore be avoided. Rock lobster and spiny lobster are sustainably fished if they are from the United States or Australia, but otherwise are probably not. There is, as we've seen, an ongoing debate about the environmental sustainability of the Chesapeake Bay blue crab fishery, but it is better regulated than many Asian crab fisheries, where imports originate. In some areas, mollusks like scallops and oysters are obtained by dredging the ocean bottom, which causes serious damage and catches many unwanted fish and other creatures. But scallops, oysters, and mussels are also farmed sustainably on ropes suspended in the sea, so removing them does not disturb the sea bottom. On the question of pain, for the reasons given in Chapter 9, we think crustacea like squid, lobster, crabs, and shrimp—along with at least one mollusk, the octopus—should be regarded as capable of feeling pain. It is much less likely, however, that bivalves like clams, scallops, oysters, and mussels could experience pain. So when these shellfish are sustainably produced, there is no strong ethical reason against eating them.

## ORGANIC, LOCAL, AND FAIR TRADE

**Organic Food:** Organic labeling schemes are never perfect. Making the standards for organic certification easy to apply in a consistent manner involves taking a philosophy of agriculture, or even a way of life, and reducing it to a checklist of points that can be verified by inspectors. As we saw in Chapter 14, big corporations have gone into organic farming and pushed the rules to their limits. Some small farmers may be following the spirit of organic farming better than these big corporations, but for various reasons, including the cost of certification, their products may not carry an "organic" label. Nevertheless, in most cases buying organic means less chemical fertilizer runoff, fewer herbicides and pesticides in the environment, more birds and animals around the farm, better soil conservation, and, in the long run, sustainable productivity.

Organically-grown plants are not genetically modified, so they pose no risk of gene flow into wild plants that could disturb natural ecosystems. The welfare of animals used in organic agriculture will be at least somewhat better than those kept in conventional factory-farms, although on some big organic farms the difference may be marginal. On the other hand, organically-produced food is more expensive than conventional food. While it would, as Cyd Szymanski suggested, be wrong to complain that organic food is too expensive and then buy a $4.50 latté, it is harder to object to buying conventionally-produced food and donating the savings to fighting global poverty. But voting with your dollars and supporting more environmentally-friendly agriculture is important, and for those who can afford it, organic is a good choice.

**Local Food:** There are various reasons why, other things being equal, it is better to buy local food. The most important is reducing the use of fossil fuels. Greater transparency is another. But other issues arise. Some of these relate directly to energy usage:

- Local early vegetables may have been grown with heat, using more fuel than required to transport them from a warmer growing region.
- Delivering small quantities of local products to many different markets may use more fuel than trucking a full load to a more distant supermarket.
- Consumers who drive to outlying local farms or markets instead of doing one-stop shopping at a supermarket may use as much fuel as

would have gone into bringing the products from more distant growers to their supermarket.

- Food production in another country may be less energy intensive than domestic production, and the difference may be greater than the energy used in shipping the food thousands of miles.

Before buying locally, we should also consider the benefits that, as we saw in Chapter 11, trade brings to farmers in distant countries who are much poorer than our local farmers. What all this suggests is that the recommendation "Buy Local!" is too simple a principle to provide sound ethical guidance. The most one can say is: "Buying local food, when it is in season, is generally a good thing to do, but sometimes there are stronger ethical reasons for buying imported food."

**Fair Trade:** Fair trade schemes ensure that more of your money gets to the people who actually grow your food; the higher earnings benefit their communities and encourage sustainable farming methods. Choose fair trade products when you buy coffee, tea, chocolate, bananas, and other items for which there are fair trade brands in your store. If your store doesn't stock fair trade items and you know that they are available, ask the store to stock them. (In the U.K., you can find out where to buy fair trade products at www.fairtrade.org.uk, or purchase them online, at www.ethicalfoods.co.uk

## HUMANELY RAISED, VEGETARIAN, OR VEGAN?

We have argued that ethical consumers will avoid factory-farmed products and most seafood. In Chapter 17 we examined the view that it is ethical to eat animals who have lived good lives and would not have existed at all, except for the ready market for their meat. We found that in practice it is often difficult to determine when animals have truly been well-treated. We also expressed our concern that treating animals as commodities for sale leads producers to seek to maximize their profits in ways that are contrary to the interests of the animals. Some animal welfare advocates have instituted labeling schemes to try to ensure that this does not happen. In the United States, Humane Farm Animal Care authorizes those producers who meet its standards to use its "Certified Humane" seal. In Britain, the RSPCA offers a highly successful "Freedom

Food" program, and other RSPCA accreditation schemes exist in Australia, New Zealand, and British Columbia. In Europe, there are the French "Label Rouge" program and the Austrian "Animal Index."

There are compromises, however, in setting the standards the farms must meet for certification, and one can also question how well the standards are met when inspectors are not present. Even when the animals from whom animal products are obtained live reasonably well, there may be suffering elsewhere in the production chain. Chickens raised outdoors for meat usually come from breeds selected to grow very quickly, and the breeder birds, their parents, will have been kept permanently hungry to prevent them from becoming too obese to breed. It is a rare dairy farmer indeed who does not separate calves from their mothers soon after birth—and what is the fate of the male calves then? And what about slaughter? Certification schemes generally are silent on that topic, so meat may be labeled "humanely raised" but may not have been humanely killed. Those who eat this meat have a duty to inform themselves about what that can mean for the animal.

If, as seems likely, going vegan is still too big a step for most of the hundreds of millions of people in industrialized countries who now eat animals, we urgently need another alternative to eating factory farmed products. A commercially successful, animal-friendly, environmentally sustainable form of agriculture looks like a promising option. Truly conscientious omnivores, however, are going to have to put time and effort into finding farms that are genuinely humane. "Certified Humane" and similar certification schemes help. Products with such labels are usually preferable to similar uncertified animal products, but it is questionable whether their standards are high enough. One possible moral rule would be: Only buy animal products if you have visited the farm from which they came—and observed procedures like searing off the beaks of laying hens, if the farm has debeaked hens. Faced with that rule, many people would find it simpler to avoid eating animals altogether. A few might put in the effort required to find farmers who show real concern for their animals and maintain the highest standards of animal welfare.

Suppose, however, that you object to the idea of killing young, healthy animals so you can eat them. That ethical view leads many people to become vegetarian, while continuing to eat eggs and dairy products. But it is not possible to produce laying hens without also producing male chickens, and since these male chicks have no commercial value, they are invariably killed as soon as they have been sexed. The laying hens themselves will be killed once their rate of laying declines. In the dairy industry much the same thing happens—the male calves are killed immediately or raised for veal, and the cows are turned into hamburger long before normal old age. So rejecting the killing of animals points to a vegan, rather than a vegetarian, diet.

Becoming a vegan is a sure way of completely avoiding participation in the abuse of farm animals. Vegans are living demonstrations of the fact that we do not need to exploit animals for food. The vegan diet is also environmentally friendly (although not more so than a diet that includes some organic animal products from animals grazing in a sustainable way on pasture that is unsuitable for growing crops). And there are now so many substitutes for animal products that becoming a vegan is far easier than it has ever been. Soy milk, along with rice milk, is now available almost everywhere, and soy yoghurt is also popular. For those who, like Mary Ann Masarech, still crave bacon, there are excellent vegan bacon substitutes, usually available in the refrigerated section of supermarkets or natural food stores There are also vegan sausages, in many different varieties, along with other vegetable-based meat substitutes. And, of course, there are many vegan cookbooks packed with an endless variety of recipes drawn from cuisines as diverse as Japan, China, Thailand, India, the Middle East, and Italy.

## THE ETHICS OF OBESITY

"Waste not, want not," the old saying went. We have already seen how much food people in industrialized nations waste. The proportion of food wasted in America would rise much higher still, however, if we included not just what people put in their garbage, but also what they put in their mouths above and beyond reasonable nutritional requirements. Eating too much should be seen not only as a health problem, but also as an ethical issue, because it wastes limited resources, adds to pollution, and increases animal suffering. The average American today eats 64 pounds more meat, poultry, and fish a year than his or her counterpart in the 1950s.[2] That's almost a 50 percent increase—and Americans were not undernourished then. As a result, 3 out of 10 Americans are now obese. In Britain, the figure is more like 1 in 5.[3] This has ethical implications.

If I choose to overeat and develop obesity-related health problems that require medical care, other people will probably have to bear some of the cost—through increased taxes needed to support my healthcare or through higher insurance premiums. A recent study in *Health Affairs* reported that overweight people with private health insurance incur insurance outlays that are, on average, $1,200 more per person than people with healthy weights.[4] The U.S. Centers for Disease Control reports that the annual medical costs attributable to overweight and obesity add $20 to $28 billion to private insurance bills and $25 to $38 billion to taxes.[5] That's an average annual cost of up to $300 for every American adult. Choosing an unhealthy diet may seem like a personal choice, but it's not fair to the people who ultimately have to pay for it. If Americans were to cut back to the meat-eating levels of the 1950s, that would improve health and slash health care costs. It would also reduce the number of animals suffering on factory farms by about the same amount as if roughly 80 million Americans became vegans.

The sin relevant to food on which Christianity has placed most emphasis is gluttony. Yet despite the strong Christian influence on American life and culture, it is hard to see much Christian influence in that country on attitudes to over-eating. It would be wrong to say that everyone who is obese commits this sin. Some have eating disorders or metabolic problems that are difficult to control. But others just eat too much and should show more restraint. Along with the old-fashioned virtue of frugality, the idea that it is wrong to be a glutton is in urgent need of revival.

## FOOD IS AN ETHICAL ISSUE—BUT YOU DON'T HAVE TO BE FANATICAL ABOUT IT

Sometimes the very success of the ethical consumer movement and the proliferation of consumer concerns it has spawned seems to threaten the entire ethical consumption project. When one ethical concern is heaped upon another and we struggle to be sure that our purchases do not contribute to slave labor, animal exploitation, land degradation, wetland pollution, rural depopulation, unfair trade practices, global warming, and the destruction of rainforests, it may all seem so complicated that we could be tempted to forget about everything except eating what we like and can afford.

When we feel overwhelmed, it is important to avoid the mistake of thinking that if you have ethical reasons for doing something, you have to do it all the time, no matter what. Some religions, like Orthodox Judaism, Islam, and Hinduism, have strict rules against eating particular foods, and their adherents are supposed to follow these rules all the time. If they break them they may feel polluted, or disobedient to their god. But this rule-based view isn't the only possible approach to ethics, nor the best one, in our view. Ethical thinking can be sensitive to circumstances.

Suppose that your elderly uncle Bob lives in a town two hundred miles from where you live, and you are his only living relative. Is it wrong not to go to see him on his birthday? The answer might well depend on how much he would enjoy seeing you, whether you have a car, or, if not, whether you can get there by bus, whether you can easily afford the bus fare, what else you could do with your time, and so on. In thinking about these things, you are paying attention to the consequences of what you are doing. How much of a difference will my visit make to Bob? How much of a sacrifice, for me or others, does it require? Similarly, a sound ethical approach to food will ask: what difference does it make, if I eat this food? How do my food choices affect myself and others? It's not wrong, in answering that question, to give some weight to your own interests and even your own convenience, as long as you don't do it to a degree that outweighs the major interests of others affected by your choices. You can be ethical without being fanatical.

Amanda Paulson, writing in the *Christian Science Monitor* about "One woman's quest to enjoy her dinner without guilt," describes the ethics of Daren Firestone, a Chicago law student who won't buy meat, but will eat the remnants of a big Thanksgiving dinner before they get tossed out. Whether or not you agree with that view—don't eat meat unless it will otherwise be wasted—there is nothing that disqualifies it as an ethical principle. Yale philosophy professor Shelly Kagan takes the same view about airline meals. A vegetarian in his everyday life, he orders meatless meals when he flies. Airlines, however, sometimes fail to deliver on such requests. If that happens, and he is offered a meat meal that he knows will be thrown out if he doesn't eat it, he'll eat it. In these circumstances—in contrast to buying meat at the supermarket—his consumption of meat seems to make no difference to the demand for it. (It's like dumpster diving without getting your clothes dirty.) Nevertheless, by not making a fuss, Kagan is sending the airlines a message that failing to provide a vegetarian meal is not a serious problem. He might also be missing an opportunity to start a conversation with the passenger next to him about why he is a vegetarian.

We are not too concerned about trivial infractions of the ethical guidelines we have suggested. We think intensive dairy production is unethical. Because dairy products are in so many foods, avoiding them entirely can make life difficult. But remember, eating ethically doesn't have to be like keeping kosher. You can take into account how difficult it is to avoid factory-farmed dairy products, and how much support you would be giving to the dairy industry if you were to buy an energy bar that includes a trace of skim milk powder. Personal purity isn't really the issue. Not supporting animal abuse—and persuading others not to support it—is. Giving people the impression that it is virtually impossible to be vegan doesn't help animals at all.

How relaxed can we be? Firestone's dietary rules also include what she calls "the Paris exemption:" if she is lucky enough to find herself in a fine restaurant in Paris—or, very occasionally, in a truly outstanding restaurant elsewhere—she allows herself to eat whatever she likes.[6] We wondered whether she believes that on these rare occasions, the pleasure that she gets from eating meat outweighs the contribution her meal makes to animal suffering. When we contacted her, however, she readily admitted that her "Paris exemption" is "more self-indulgence than utilitarian calculus." But that doesn't mean that her general opposition to eating meat is not ethical. It is, but she gives more weight to what she wants to do than she would if she were acting on strictly ethical principles all the time. Very few of us are in any position to criticize that, and most of those who do criticize it are deceiving themselves about their choices when their own desires are at stake. A little self-indulgence, if you can keep it under firm control, doesn't make you a moral monster, and it certainly doesn't mean that you might as well abandon your principles entirely. In fact, Firestone believes that by allowing herself to satisfy her occasional cravings—maybe once every three months—she has been able to be faithful to her principles for many years, while other vegetarians she knows have given up the whole practice because one day they could not resist the smell of bacon frying.

At the opposite end to the "Paris exemption" is the "hardship exemption." Factory farming and other unethical methods of producing food have spread because they lead to food that sells for less than food produced by more traditional methods. Replacing these foods with organically produced food generally means paying more for your food. In recommending foods that are more expensive, we are not saying that people who can comfortably afford organic food and humanely-raised meat are more ethical than those who cannot. As with your visit to uncle Bob, circumstances matter. If Bob would very much like to see you, but the bus fare will make such a dent in your budget that your children will go to bed hungry, no one will blame you for not going. But few families in industrialized nations are as poor as that. In the United States, families that consider themselves poor often drink sodas rather than water. Shopping at Wal-Mart, Jake bought a lot of prepared, packaged food like corn dogs, steak fingers, and breaded fish fillets. Unfortunately, these products deliver poor value for money in terms of their nutritional content. Food that is both more ethical and more economical is available in every supermarket. Buying organic food without incurring extra expense, on the other hand, is usually not possible. Taking that into account, and considering that there are more powerful grounds for avoiding factory-farmed products than for buying only organic food, it is reasonable to limit the obligation to buy organic food to what one can afford without undue hardship, while seeing the obligation to avoid factory farm products as more stringent.

No other human activity has had as great an impact on our planet as agriculture. When we buy food we are taking part in a vast global industry. Americans spend more than a trillion dollars on food every year. That's more than double what they spend on motor vehicles, and also more than double what the government spends on defense. We are all consumers of food, and we are all affected to some degree by the pollution that the food industry produces. In addition to its impact on over six billion humans, the food industry also directly affects more than fifty billion nonhuman land animals a year.[7] For many of them, it controls almost every aspect of their lives, causing them to be brought into existence, reared in totally artificial, factory-style production units and then slaughtered. Additional billions of fish and other sea creatures are swept up out of the sea and killed so we can eat them. Through the chemicals and hormones it puts into the rivers and seas and the spread of diseases like avian influenza, agriculture indirectly affects all living creatures. All of this happens because of our choices about what we eat. We can make better choices.

# WHERE TO FIND ETHICAL FOOD

Wherever you buy food—whether at a supermarket, natural food store, farmers' market, or restaurant—talk to the managers about the ethical concerns most important to you. Tell them about the five ethical principles shown on pages 270–71 of this book. Ask them for information about the sources and production methods of the foods that you usually buy. Ask them to stock some of the products or brands that you know meet your ethical standards, and then, of course, be sure to go back and buy them. This will make ethical foods more available to other consumers, so your request is an ethical act in itself.

## IN GENERAL:

We wrote about Whole Foods Market in Chapter 12. Whole Foods has acquired the UK organic food retailer Fresh and Wild, with stores in London and Bristol, and is planning to expand. For store locations, see here: http://wholefoodsmarket.com/strores/list_freshandwild.html. Planet organic has three stores in London. For details, see www.planetorganic.com.

The Animal Free Shopper is a popular resource containing details of vegan products available in the U.K., including food and drink, footwear and clothing, toiletries and cosmetics, baby and child care and household products: www.animalfreeshopper.com/.

The Guide to Compassionate Food Shopping contains information on the availability of high-welfare animal products, and is available to download from www.ciwf.org.uk/publications/consumers.html.

www.farmgatedirect.com is an online shop selling RSPCA Freedom Food scheme certified products, ensuring better standards of animal welfare. They deliver across the U.K.

The Good Shopping Guide is a very comprehensive ethical shopping reference book. It can be ordered from www.ethical-company-organisation.org , or from bookshops.

www.gooshing.co.uk is a free ethical shopping tool that enables free ethical comparisons of over 250,000 products, including food and drink. The Ethical Consumer Magazine has an online shopper's guide that provides an 'ethical score' for a range of products and services: www.ethiscore.org. Alternatively, look for a health/natural food store near you at www.greenpeople.org

More and more restaurants are including vegetarian, organic, sustainable, and other ethical choices on their menus. Look for these at the places where you frequently dine. If none are offered, ask the management to offer them. If they won't accommodate you, check the restaurant listings in your phone book and phone them to ask what they offer. Just making the calls helps to alert them that providing ethical food choices will attract customers. You may also be able to find restaurants that offer vegetarian food choices in the 'Vegetarian Britain' guide, featuring vegetarian restaurants and take-away across Britain, vegetarian-friendly places to stay, as well as health food and whole food stores: www.vegetarianguides.co.uk/

## ANIMALS AND THE ENVIRONMENT

The surest way to avoid harming animals and the environment is to avoid animal products altogether. To get started, you can get advice and recipes from Vegan Outreach at www.veganoutreach.org. People for the Ethical Treatment of Animals offers two helpful Web sites:

www.GoVeg.com and www.VegCooking.com. In the UK, go to the website of the original Vegan Society, at www.vegansociety.com, or to the even older Vegetarian Society at www.vegsoc.org. Other useful sources of information are www.veganvillage.co.uk and the website of Animal Aid, at www.animalaid.org.uk/veggie.

If you feel that you can't go all the way, all at once, at least try vegan meals some days each week. You may already be doing this unintentionally if you choose some of the offerings at Italian, Chinese, Middle Eastern, or Indian restaurants. It's really not that difficult, and you'll like the food.

You could also join others in Meatless Mondays, sponsored by the Johns Hopkins Bloomberg School of Public Health and the Center for a Livable Future at www.meatlessmonday.com. Another option is to follow the three R's—Reduce, Refine, Replace—recommended by the Humane Society of the United States at www.humaneeating.org. Alternatively visit www.eatlessmeat.org

Above all, avoid animal products that come from factory farms. See "Local Farms" below. A useful Web site for those interested in more compassionate eating, especially in the U.S. and Canada but with links to some other countries, is http://eatkind.net. For online order of ethical foods, try www.ethicalfoods.co.uk.

## FAIR TRADE

Information about fair-traded foods and suppliers is available from www.fairtrade.org.uk, Fair Trade Labeling Organizations International at www.fairtrade.net, and the Fair Trade Federation at www.fairtrade-federation.org. In addition www.traidcraftshop.co.uk is a fair-trade online shop and has a wide range of products, from food to clothing and decorations.

## LOCAL FARMS

Whenever you can, buy from local farmers who are willing to talk to you about your concerns and to show you their methods and systems. Information about local farms and farmers' markets, including location and dates, farm shops, pick-your-own farms and local producers across the

U.K., is available at www.farmersmarkets.net and www.farma.org.uk. Local may not always mean 'welfare-friendly'—so do ask.

## ORGANIC

You can find organic foods in food co-ops, health/natural food stores, and most supermarkets. Ask local farmers and those selling at farmers' markets. For a resource on everything organic, including information on nutrition, organic festivals and a local food guide, try www.soilassociation.org. For information on where to buy organic food in the U.K., including a searchable directory of local organic box schemes and local shops that sell organic, go to www.organicfood.co.uk/ or to www.whyorganic.org/ which offers a searchable organic directory, listing organic retailers, box schemes, farm shops, manufacturers, restaurants and accommodation.

## SUSTAINABLE FISH

If you eat marine animals, choose species that come from sustainable fisheries. When shopping or dining out, use the Seafood Alliance choices, listed at www.seafoodchoices.org.

The Marine Stewardship Council runs a labelling scheme that ensures that wild fish come from a well-managed fishery that does not contribute to over fishing (but does not include welfare standards). At http://eng.msc.org/ you can find a list of products available with the MSC label in the U.K. There are serious concerns about the ethics of farming fish, but if you do buy farmed fish, such as salmon and trout, look out for organic or the RSPCA's Freedom Food scheme—these ensure better standards of fish welfare in the U.K.

# WHERE TO FIND INFORMATION

To learn more about the issues discussed in this book:

## FOOD AND AGRICULTURE IN GENERAL

Eric Schlosser's best seller, *Fast Food Nation: The Dark Side of the All-American Meal* (New York: Perennial, 2002), is still the best-told story of America's fast food industry and its influence on eating habits.

Ann Cooper, *Bitter Harvest: A Chef's Perspective on the Hidden Dangers in the Foods We Eat and What You Can Do About It* (New York: Routledge, 2000). The book is weighted toward food safety and quality, but it contains a good history and other interesting information about our industrialized food and agricultural system.

Marion Nestle, *Food Politics: How the Food Industry Influences Nutrition and Health* (Berkeley: University of California Press, 2002). The title tells it all: Corporate power, lobbyists, advertising, and marketing tricks.

A UK organization, Ethical Consumer Research Association (ECRA), offers information to help consumers make ethical choices of food and products: www.ethicalconsumer.org. The materials on fair trade, GMOs, and other concerns should be helpful to consumers everywhere.

Other useful titles include: Tim Lang and Michael Heasman, *Food Wars: The Global Battle for Mouths, Minds and Markets* (London: Earthscan, 2004). Colin Tudge, *So Shall We Reap: What's Gone Wrong with the World's Food and How to Fix it* (London: Penguin, 2004). *The Global Benefits of Eating Less Meat*, CIWF Trust, 2004: available to download from: www.ciwf.org.uk/publications and in the index, select 'farming and the environment'. John Webster, *Animal Welfare: Limping Towards Eden* (Oxford: Blackwell, 2005). Jacky Turner and Joyce D'Silva (eds), *Animals, Ethics and Trade: The challenge of Animal Sentience* (London: Earthscan, 2006).

Two recent books that are well worth reading are:

Marion Nestle, *What to Eat* (North Point Press, New York, 2006). The well-known nutritionist takes us through the supermarket, offering advice on what to eat.

Michael Pollan, *The Omnivore's Dilemma: A Natural History of Four Meals* (Penguin, New York, 2006). We have discussed some of Pollan's writing in the preceding pages. This is his fullest statement on our food choices.

## ANIMAL AGRICULTURE

Jim Mason and Peter Singer, *Animal Factories: What Agribusiness Is Doing to the Family Farm, the Environment and Your Health* (New York: Harmony Books, rev. ed. 1990). Although now out of print (copies are still available in libraries and on used-book Web sites) and the figures dated, the book explores all of the ills of factory farming; with photographs of factory farms, inside and out. The factory farms of today are essentially the same, only much larger.

Jim Mason, *An Unnatural Order: The Roots of Our Destruction of Nature* (New York: Lantern Books, 2005). The book examines ideas about animals, food, and gender from pre-historic times—including scavenging, hunting, animal domestication, animal sacrifice, and the influence of animal agriculture on the shaping of early Western civilizations.

Karen Davis, *Prisoned Chickens, Poisoned Eggs: An Inside Look at the Modern Poultry Industry* (Summertown, TN: Book Publishing Company, 1996). Well-researched details on modern chicken and egg factories.

Karen Davis, *More Than a Meal: The Turkey in History, Myth, Ritual, and Reality* (New York: Lantern Books, 2001. The same excellent research on the only native American—and unfairly maligned—farmed animal.

Danielle Nierenberg, *Happier Meals: Rethinking the Global Meat Industry*, Worldwatch paper 171 (Washington, DC: Worldwatch Institute, 2005). A look at the worldwide growth of meat consumption and factory farming methods; with suggestions for alternatives.

Erik Marcus, *Meat Market: Animals, Ethics, and Money*, Brio Press, Ithaca, NY 2005. Marcus carefully reviews current animal agricultural practices and data, and advocates new strategies for reducing the suffering of farmed animals.

Ken Midkiff, *The Meat You Eat: How Corporate Farming Has Endangered America's Food Supply,* New York: St Martin's Press, 2004 . . . as well as our health, food quality and diversity, the environment, and farmed animals.

Bernard E. Rollin, *Farm Animal Welfare: Social, Bioethical, and Research Issues*, Iowa State Press 1995. The author, University Distinguished Professor, Professor of Philosophy, Professor of Biomedical Sciences, Professor of Animal Sciences, and University Bioethicist at Colorado State University, reviews the science relevant to the most controversial animal farming practices.

Up-to-date information about factory farms is available from Farm Sanctuary at www.factoryfarming.com and Global Resource Action Center for the Environment (GRACE) at www.factoryfarming.org.

Compassion in World Farming, based in Britain but with affiliates in several other countries, has many thoroughly researched publications on farming, food, and fish. See: www.ciwf.org.uk/publications. This includes a website on animal sentience www.animalsentience.com, and resources on Good Agricultural Practice from a farm animal welfare perspective http://www.ciwf.org.uk/education/gaphome.html.

To see factory farm conditions for yourself, watch the video "Meet Your Meat" narrated by actor and activist Alec Baldwin. www.petatv.

com/tvpopup/Prefs.asp?video=mym2002. Animal Aid (www.animalaid.org.uk) has various resources on issues such as the dangers of factory farming and the suffering of farmed animals.

To see what Tyson Foods, Inc. has to say, go to: www.tyson.com.

McDonald's on corporate responsibility: www.mcdonalds.com/corp/values/socialrespons.html.

## ENVIRONMENT

Jim Motavalli edits *E, The Environmental Magazine*, which regularly runs articles on agriculture, fisheries, fair trade and other geo-political concerns. You may find information on issues of concern to you in *E*'s back issues and archives. www.emagazine.com

*Green Living: The E Magazine Handbook for Living Lightly on the Earth*. By the editors of *E, The Environmental Magazine*, 28 Knight Street, Norwalk, CT 06851.

Sierra Club catalogs environmental misdeeds of factory farms in "The Rapsheet on Animal Factories" and has campaigns. See: www.sierraclub.org/factoryfarms.

Waterkeeper Alliance's Robert F. Kennedy, Jr., is America's most visible and vocal critic of animal factories. See: www.waterkeeper.org.

Carolyn Johnsen, *Raising a Stink: The Struggle Over Factory Hog Farms in Nebraska*, University of Nebraska Press, Lincoln, 2003. A balanced report on the complex and deeply divisive conflicts among factory farmers, local residents, family farmers, environmentalists, and elected officials.

"Watching What We Eat" is a useful article by Brian Halweil and Danielle Nierenberg on agriculture's impacts. In Worldwatch Institute, *State of the World 2004*, W.W. Norton & Company, 2004. Friends of the Earth (www.foe.co.uk) run a RealFood campaign, with emphasis on GM crops and the environment.

## FAIR TRADE

There are several fair trade organizations with varying missions:

Fairtrade Labelling Organization International: www.fairtrade.net

Transfair USA: www.transfairusa.org.

Fair Trade Federation: www.fairtradefederation.org

Fairtrade Foundation: www.fairtrade.org.uk

Fair Trade Resource Network: www.fairtraderesource.org

Global Exchange, an international human rights organization, promotes social, economic and environmental justice around the world. www.globalexchange.org.

For a technical review of how agricultural trade can benefit the developing world, see: M. Ataman Aksoy and John Beghin, eds., *Global Agricultural Trade and Developing Countries*, World Bank, Washington, DC, 2005.

Institute for Agriculture and Trade Policy: www.iatp.org. Based in the U.S. and Switzerland, IATP works on the effects of globalization and the challenges facing local communities.

Oxfam International, *Rigged Rules and Double Standards: Trade, Globalization and the Fight Against Poverty*, Oxfam, 2002. At: www.maketradefair.com/assets/english/report_english.pdf. Information about fair trade food is available from The British Association for Fair Trade Shops, www.bafts.org.uk.

## FISHERIES AND AQUACULTURE

People for the Ethical Treatment of Animals has an overview of fishing at: www.fishinghurts.com/fishing101.asp.

The Worldwatch Institute has several reports on fish farming. See www.worldwatch.org/pubs/goodstuff/shrimp/ and www.worldwatch.org/pubs/mag/2003/165.

Colin Woodward, *Ocean's End: Travels Through Endangered Seas* (New York: Basic Books, 2000). Drawing on his travels across six continents and 100,000 miles, the author explains how pollution, harmful fishing practices, ignorance, and global warming are destroying the world's oceans.

Philip Lymbery, "In Too Deep—The Welfare of Intensively Farmed Fish," Compassion in World Farming Trust, available from www.ciwf.org.uk/publications/fish.html.

Pew Oceans Commission, *America's Living Oceans: Charting a*

*Course for Sea Change* (Arlington, VA: Pew Oceans Commission, 2003). Available at www.pewoceans.org.

Marine Stewardship Council has information about its "Fish Forever" certification of sustainable fisheries at: www.msc.org.

Environmental Defense maintains the Oceans Alive program. See their video that takes you underwater to view sea creatures at www.oceansalive.org.

## GENETICALLY MODIFIED FOODS

Lee Silver, *Challenging Nature: The Clash of Science and Spirituality at the New Frontiers of Life*. Ecco, New York, 2006. Much broader than just GMOs, this book is definitely on the side of science.

Michael Ruse and David Castle, eds., *Genetically Modified Foods: Debating Biotechnology*. Prometheus, Amherst, NY, 2002. Those for and against GM foods get to state their case here.

Food and Agriculture Organization of the United Nations, *The State of Food and Agriculture, 2003-4*, Rome, 2004. A balanced view of the global risks and benefits of GMO foods.

Union of Concerned Scientists at www.ucsusa.org; see "Food and Environment" for reports on genetically engineered foods.

Center for Food Safety on GMOs: www.centerforfoodsafety.org/geneticall7.cfm.

Pew Institute on Food and Biotechnology, http://pewagbiology.org.

*The Gene and the Stable Door: Biotechnology and Farm Animals*, CIWF Trust, 2002. Available to download from: www.ciwf.org.uk/publications.

## LOCAL FARMING AND SUSTAINABLE AGRICULTURE

Brian Halweil, *Eat Here: Homegrown Pleasures in a Global Supermarket,* Norton, New York, 2004. Explains why it is good to "eat local."

Rich Pirog, et al, *Food, Fuel and, Freeways,* Leopold Center for Sustainable Agriculture, University of Iowa, Ames, Iowa, 2001. Available at www.leopold.iastate.edu/pubs/staff/ppp/food_mil.pdf.

Anna Lappe and Bryant Terry, *Grub Ideas for an Urban Organic Kitchen*. Tarcher/Penguin Group, 2006. Recipes, meals oriented around the seasons, resource lists, graphics, and do-it- yourself tip sheets, charts, and checklists.

Food Routes is a non-profit organization for the "buy local" movement: www.foodroutes.org; Local Harvest is another such organization: www.localharvest.org.

The National Campaign for Sustainable Agriculture works with over a hundred partner organizations to a more fair and environmentally sustainable U.S. agricultural policy. www.sustainableagriculture.net.

Sustain: the alliance for better food and farming (www.sustainweb.org) advocates food and agriculture practices that enhance the health and welfare of people and animals.

## ORGANIC FARMING

Rodale Institute, the granddaddy of the organic movement, has many publications: www.rodaleinstitute.org and www.newfarm.org.

Julie Guthman, *Agrarian Dreams: The Paradox of Organic Farming in California*, University of California Press. August, 2004. A study of the development of organic farming in California, revealing some of the differences and similarities between organic and conventional farming.

Michael Sligh and Caroline Christman, *Who Owns Organic? The Global Status, Prospects and Challenges of a Changing Organic Market,* Rural Advancement Foundation International—USA, Pittsboro, NC, 2003.

The Soil Association (www.soilassociation.org) is Britain's leading organic body and has a wealth of downloadable publications on organic farming.

## SLAUGHTERHOUSE WORKERS

Gail A. Eisnitz, *Slaughterhouse: The Shocking Story of Greed, Neglect, and Inhumane Treatment Inside the U.S. Meat Industry* (Amherst, NY: Prometheus Books, 1997). A fast-paced, readable account of what it's like inside the tough slaughterhouse sub-culture and how meat industry pressure has rendered the USDA inspection system ineffective.

Donald D. Stull and Michael J. Broadway, *Slaughterhouse Blues: The Meat and Poultry Industry in North America* (Belmont, CA: Thomson/Wadsworth, 2004). Fifteen years of research by a cultural anthropologist and a social geographer on the communities surrounding slaughter plants.

Human Rights Watch, *Blood, Sweat and Fear: Workers' Rights in U.S. Meat and Poultry Plants* (New York: Human Rights Watch, 2004).

## VEGETARIANS AND VEGANS

The Vegan Society has published an excellent book on vegan nutrition: Plant Based Nutrition and Health, by Stephen Walsh. Available to buy from www.vegansociety.com , or from bookshops.

The Vegetarian Society produces information sheets on various issues such as vegetarian nutrition and the history of the vegetarian movement—these are available on their website www.vegsoc.org. They also have a wide range of vegetarian and vegan recipes, and a database of approved products.

Animal Aid provide useful vegan/vegetarian resources, including a guide to going dairy free, and a free veggie starter pack for UK residents. www.animalaid.org.uk

Vegetarian Resource Group has solidly researched materials: www.vrg.org.

Erik Marcus, noted above in "Animal Agriculture," is a respected writer on vegan issues: www.vegan.com.

Try the monthly magazine *VegNews* for comprehensive coverage of news, nutrition, lifestyles, book reviews and recipes: www.vegnews.com.

JoAnn Farb's book, *Compassionate Souls: Raising the Next Generation to Change the World,* is available from Lantern Books, New York. www.lanternbooks.com

## TO REACH BUSINESSES DISCUSSED IN THIS BOOK

Chipotle: www.chipotle.com
McDonalds: www.mcdonalds.com
Niman Ranch: www.nimanranch.com
Tyson Foods: www.tyson.com
Wal-Mart: www.walmart.com
White Dog Café: www.whitedogcafe.com
Whole Foods supermarkets: www.wholefoodsmarket.com
Wild Oats: www.wildoats.com

# ACKNOWLEDGMENTS

There are so many people to thank for their assistance with this book that despite our best efforts, we will probably forget to thank some. To them, we apologize in advance. But at least we know where to begin: with our three families, Jake Hillard and Lee Nierstheimer, Jim Motavalli and Mary Ann Masarech, and Joe and JoAnn Farb. They let us into their homes and, despite their busy lives, gave us generous amounts of their time over the past two years, sharing shopping lists, answering questions, checking and double-checking labels, and cooperating with us every step of the way as we researched and wrote this book. Two other families do not appear in this book, but also gave their time and shared their food choices in the early stages of this project, before we reluctantly decided that tracing the choices of five families would make the book too unwieldy. They are Mark Barr and Rebecca Adamson in Kansas City, Missouri, and Jerry and Joanna Shelton, in Minneapolis, Minnesota. We thank all of you for making this book possible.

We had two wonderful research assistants in Mary Finelli and Gaverick Matheny. Their diligence, research skills, and extraordinary knowledge of many topics covered in this book made them invaluable and it is hard to imagine how we could have done it without them. Kim McCoy took a chaos of names, companies, brands, labels, and Web sites and converted all of it into neat, color-coded tables. Her organizing skills made it much easier for us to track correspondence and various information by company and by product. Others who assisted with research in more minor ways were Shawna Benston, Joe Corelli, and Oren Rosenbaum.

Diane and Marlene Halverson, humane animal husbandry experts with the Animal Welfare Institute, organized our visits to pasture pig farms in North Carolina and flew halfway across the country to personally introduce us to the farmers. They helped us with many questions and problems in our work on this book. Dawnwatch (www.dawnwatch.com) was a regular source of valuable information. Karen Davis, of United Poultry Concerns (www.upc-online.org) provided critical feedback on our chapters on chickens.

The fact that so few farmers were willing to show us over their farms makes us all the more grateful to the few who were: "Wayne Bradley," who we cannot thank properly because he asked us to use a pseudonym; George, Tim, and Mike Holmes, Holmes Farm, Creswell, North Carolina; Mike and Suzanne Jones, Mae Farm, Louisburg, North Carolina; Rodney and Judith Martin, Bridge View Dairy, Oxford, Pennsylvania; Kip and Jackie Glass, Autumn Olive Farms, Bois D'Arc, Missouri; Rick Hopkins and Diana Botsford, Buck Prairie Greens Farm, Marionville, Missouri; Dave and Florence Minar, Cedar Summit Farms, New Prague, Minnesota; Jesse and Sandra Laflamme of Pete and Gerry's Organic Eggs, Monroe, New Hampshire; Arden Landis, organic dairy farmer, Kirkwood, Pennsylvania; and Patrick and Anne Francis, Romsey, Victoria, Australia.

Other farmers who generously shared their time and experience with us by telephone are: Travis Forgues, Vermont organic dairy farmer; Elizabeth Henderson, Northeast Organic Farming Association of New York; Vivien Strauss, Strauss Family Creamery, Petaluma, California; Tom Sherwood and family, Sherwood Farms, Easton, Connecticut; Mark and Connie Tjelmland, TJ Family Farm, McCallsburg, Iowa; Jane Bush, Grazing Fields, Charlotte, Michigan egg farmer; and Lynn Garling, Over The Moon Farm, Rebersburg, Pennsylvania egg farmer.

We thank the following people in the food business for taking the time to talk with us about their practices and products: Bob Wills, Cedar Grove Cheese in Plain, Wisconsin; Steve McDonnell, Applegate Farms in Branchburg, New Jersey; Steve Ells, Chipotle in Denver, Colorado; Judy Wicks, White Dog Café in Philadelphia; Nancy Rosenberg, Horizon Foods in Plainview, New York; John Mackey, Whole Foods Markets in Austin, Texas; Bill Niman and Paul Willis, Niman Ranch in Oakland,

California; Doug Rauch, Trader Joe's East Coast in Needham, Massachusetts; Matt Koch, Road's End Organics in Morrisville, Vermont; Michael Potter, Eden Foods in Clinton, Michigan; George Siemon, Nick Levendoski and David Bruce, Organic Valley in La Farge, Wisconsin; Kevin Murphy, *Food Industry Insider* in Lenexa, Kansas; Cyd Szymanski, Nest Fresh Eggs in Denver, Colorado; Graeme Carrie, Frenzs, New Zealand; Bob Hodges and Ronald Bennet, Moark in Chesterfield, Missouri; Chris Henning, Wild Oats Markets in Boulder, Colorado; Walt Coleman, Coleman Dairy in Little Rock, Arkansas; Denis Hilton, Customer Relations, the Cooperative Group, UK, and Bob Langert, McDonald's Corporation.

Rich Wood, Steve Roach, and Kathy Seuss of Food Animal Concerns Trust (FACT) in Chicago shared their work on egg carton labeling, and on FDA and USDA efforts to regulate the feeding of animal wastes to ruminants. Brenda Davis, Fisheries Biologist, Maryland Department of Natural Resources, supplied information on the Chesapeake Bay crabbing industry and Jim Ianelli, Ph.D., biologist, National Oceanic and Atmospheric Administration, Alaska, did the same for Alaska's pollock fishery. Greg Bowman and Dan Sullivan, *New Farm* magazine, gave us information about organic farming and contacts to organic farmers. Dr. Temple Grandin, livestock handling expert, answered many of our questions on improvements in slaughter procedures and on other farm animal issues. Paul Shapiro and Miyun Park, formerly of Compassion Over Killing and now with the Humane Society of the United States, kept us informed on the progress of their efforts to introduce truthfulness into egg carton labeling. Kathleen Goldstein of Environmental Defense gave us information on the stocks of various species of fish and other sea creatures. Michael Shimkin, of Transfair USA, helped us find information on fair trade, and Sasha Courville, of the International Social and Environmental Accreditation and Labelling (ISEAL) Alliance briefed us on the complex world of labeling foods in accordance with their social and environmental impact. Marjorie Victor and Seth Petchers, of Oxfam America, responded to our queries on fair trade coffee. Charles Walaga, the Ugandan member of the International Federation of Organic Agriculture Movements' Development Forum, shared his views of the benefits of food exports for developing countries.

In Britain, Joyce D'Silva of Compassion in World Farming gave us the benefit of her extensive knowledge of developments in farming in Britain and the European Union, and assisted us with listing useful resources for the UK edition. Andrew Tyler of Animal Aid took the time to respond to our questions regarding providing animal foods to impoverished people in developing countries. We are grateful to John Robbins for answering some questions too. Daren Firestone told us about his dietary practices. Danielle Nierenberg, Worldwatch Institute, helped with our Resouces list. Bruce Friedrich of People for the Ethical Treatment of Animals and Karen Dawn provided information on a variety of topics and also read a draft of the manuscript and made helpful comments and suggestions. Erik Marcus's input on water use was important for that section of the book, and he too read and commented on the draft manuscript, as did Josh Greene, Agata Sagan, and Brent Howard.

Steve Pacala of Princeton University and Phil Camill of Carleton College gave us the benefit of their expertise on carbon sequestration when we looked at that topic in regard to organic farming. Alan Kolok of the University of Nebraska gave up a day to take us on a tour of some Nebraska feedlots and provided information drawn from his research on the environmental impact of steroid use in cattle. David Pimentel of Cornell University and John Robbins explained how they had reached their published conclusions on the amount of water used to produce beef. We are also grateful to Lindsay Allen of the University of California, Davis, for being willing to talk openly with us about her research, despite her prior bad experiences with the media. John Reganold, Washington State University, sent us his research comparing organic and conventional farming. Lee Silver of Princeton University, and Bob Phelps of Genethics Network, Australia, shared their expertise about GM foods. Professor Glenn Grimes, University of Missouri, informed us about trends in the U.S. pork industry. Joy Mench of the University of California, Davis, and David Fraser of the University of British Columbia, answered specific queries on animal welfare issues. Mohan Raj of the University of Bristol, was most helpful on the stunning of chickens. Virgil Butler, former Tyson employee turned whistle blower and animal rights activist told us of his experiences. Heather House, Pennsylvania Association for Sustainable Agriculture, provided information about organic dairy farms and free range egg farms.

Matt Ball and Jack Norris of Vegan Outreach, and Neal Barnard, Trulie Ankerberg-Nobis, Sarah Farr, and Amy Lanou of Physicians Committee for Responsible Medicine, were sources on vegan diet and nutrition. John Hendrickson of the Center for Integrated Agricultural Systems, University of Wisconsin, Madison, shared his knowledge of energy and food transport. Jason Furman of New York University's Wagner Graduate School of Public Service, helped us to avoid errors in our discussion of the impact of Wal-Mart's policies on taxpayers. Dr. Richard Betz, Mark Adams, and May Belle Osborne of Southwest Missouri Citizens Against Local Moark Expansion, Neosho, Missouri, told us of their encounter with one of the nation's largest egg producers. Aloma Dew, Sierra Club Associate Midwest Representative, provided information about *Sierra Club, et al vs. Tyson Foods, Inc.* Dennis Barrow, Fairfield Historical Society, talked to us about the history of Fairfield, CT. Esther Singer put us in touch with her friends, Tim, Shane, Gareth, and Danya, and they were happy to take us along one evening to see what the dumpsters were offering. Eilene Cohhn offered ideas for promotion and gave general moral support.

For comments on the ethical arguments in chapter 17, we thank faculty and visiting fellows at the University Center for Human Values at Princeton University, members of the Centre for Applied Philosophy and Public Policy at the University of Melbourne, and those who contributed to the discussion at the International Society for Utilitarian Studies Conference at Dartmouth University in August 2005. David DeGrazia of George Washington University, and Walter Sinnott-Armstrong of Dartmouth, also gave us extensive written comments.

Our agent, Kathy Robbins, was involved in the conception of this book from the start, and together with David Halpern was frequently a source of good advice. We were delighted when Kathy found an enthusiastic response for the project from Stephanie Tade, at Rodale. Stephanie helped to shape the plan for the book, but then moved on before it could be completed. Chris Potash took her place and working with him has been a real pleasure. In innumerable ways, he has helped us to get the book right. We also thank Leigh Haber and Liz Perl at Rodale, for their close involvement in the project as it progressed. Rodale's strong support for the book has been tremendously encouraging. For the U.K. edition, I thank Ravi Mirchandani for his enthusiasm for the book and its subject, and Caroline Knight and Alban Miles for their care in its production.

With such numerous and diverse sources of information, advice and comment, it is obvious that, grateful as we are for the assistance, the final responsibility for the book remains ours.

# ENDNOTES

## PREFACE

1 Senator Robert Byrd, U.S. Senate, July 9, 2001. The speech is available at www.animalsvoice.com/PAGES/writes/editorial/investigations/legis/byrd_cruelty1.html

2 Robert F. Kennedy, Jr, "Crimes Against Nature," *Rolling Stone*, December 11, 2003; www.commondreams.org/views03/1120-01.htm.

## INTRODUCTION

1 Michel Foucault, *Histoire de la Sexualité 2: L'usage des Plaisirs* (Paris: Gallimard, 1984a), as cited by Hub Zwart, "A Short History of Food Ethics," *Journal of Agricultural and Environmental Ethics* 12: 113–126, 2000.

2 Plato, *The Republic,* Book II.

3 K. Dalmeny, E. Hanna, and T. Lobstein, *Broadcasting bad health: Why food marketing to children needs to be controlled: A report by the International Association of Consumer Food Organizations for the Wold Health Organization consultation on a global strategy for diet and health*, 2003, www.foodcomm.org.uk/Broadcasting_bad_health.pdf p. 5

4 Marion Nestle, *Food Politics: How the Food Industry Influences Nutrition and Health*, University of California Press, Los Angeles and Berkeley, 2002. See also www.foodpolitics.com.

5 See www.supersizeme.com.

6 For a full account of how veal calves are kept, see Peter Singer, *Animal Liberation*, Ecco, New York, 2001 (first published 1975). On the decline in veal consumption, see USDA Economic Research Service, *Food availability spreadsheets: Beef, veal, pork, lamb and mutton, and total red meats*, 21 Dec 2005, www.ers.usda.gov/Data/FoodConsumption/spreadsheets/mtredsu.xls

7 Nanette Hanson, "Organic food sales see healthy growth," MSNBC News, December 3, 2004, http://msnbc.msn.com/id/6638417; for the EU, see "Ikea embraces organic ingredients," *Food Navigator,* July 7, 2005 www.foodnavigator.com/news/news-ng.asp?n=61239-ikea-embraces-organic.

8 The Vegetarian Society summarizes the surveys at www.vegsoc.org/info/statveg.html.

9 Humane Society of the United States, "Wild Oats and Whole Foods Show Compassion with Cage-Free Egg Policies," www.hsus.org/farm_animals/farm_animals_news/wild_oats.html; Mackey's comment is from a speech at Princeton University, November 15, 2005.

10 JoAnn Farb, interview.

11 "McDonald's going organic—enough to change the image?" *Food Navigator*, January 29, 2003, www.foodnavigator.com/news/news-ng.asp?id=45953-mcdonalds-going-organic.

12 Information from Denis Hilton, Customer Relations, The Co-operative Group, Manchester, U.K., March 2005; Rosie Murray-West, *Daily Telegraph*, December 2, 2002.

13 *Farming U.K.*, October 26, 2005, www.farminguk.com/bsp/10130/ews.asp?DBID=103-281-013-094&iPage=1&id=3679.

14 Animal Aid Factfile, www.animalaid.org.uk/campaign/vegan/poultry01.htm citing figures from the Department of Environment, Food and Rural Affairs.

15 Robert Verkaik, "Archbishop tells Church to help save the planet with green policies," *The Independent*, February 3, 2005.

16 Steve Kopperud, "Sitting on our hands won't help," *Florida Agriculture,* April 2003, www.floridafarmbureau.org/flag/april2k3/viewapr.html.

17 Charlie Arnot, "Producers Tell Story," *Feedstuffs*, January 17, 2005.

18 Mike Owens, "Corporate Hog Farms," KSDK News, July 12, 2005, http://ksdk.com/news/news_article.aspx?storyid=81786.

19 Kevin Murphy, interview, February 17, 2005, and subsequent email.

20 Peter Cheeke, *Contemporary Issues in Animal Agriculture*, Pearson, Upper Saddle River, NJ, 3rd ed., 2004, p. 332.

21 Christy Pitney, "Gotta Believe: Food Fuels Emotion-Based Ideologies," *Food Systems Insider*, March 1, 2005, www.vancepublishing.com/FSI/articles/0503/0503believe.htm.

## CHAPTER 2

1 National Farmers Union and British Poultry Council, *British Chicken: What Price?* 2006, www.thepoultrysite.com/FeaturedArticle/FATopic.asp?Display=587.

2 Center for Nutrition Policy and Information, U.S. Department of Agriculture, *Nutrient Content of the U.S. Food Supply*, Home Economics Research Report No. 56, November 2004, p. 14. www.cnpp.usda.gov/Pubs/Food%20Supply/FoodSupply2003Rpt/FoodSupply1909-2000.pdf

3 "Tyson Today," Tyson Foods Web site, www.tyson.com.

4 "Tyson beefs up ingredients market," *Food Navigator*, July 3, 2003, www.foodnavigator.com/news/ng.asp?id=47010-tyson-beefs-up.

5 Bill Roenigk, phone conversation with Gaverick Matheny, February 23, 2004.

6 Jennifer Viegas, "Study: Chickens Think About Future," *Discovery News*, July 14, 2005, http://dsc.discovery.com/news/briefs/20050711/chicken.html.

7 Susan Milius, "The science of eeeeek: what a squeak can tell researchers about life, society, and all that," *Science News*, Sept 12, 1998; available at www.findarticles.com/p/articles/mi_m1200/is_n11_v154/ai_21156998.

8 Lesley Rogers, *The Development of Brain and Behaviour in the Chicken*, CABI Publishing, Wallingford, Oxfordshire, 1995, p. 217.

9 T. C. Danbury, et al., "Self-selection of the analgesic drug carprofen by lame broiler chickens," *Veterinary Record*, 146 (March 11, 2000) pp. 307–11.

10 Jonathan Leake, "The Secret Lives of Moody Cows," *Sunday Times*, February 27, 2005.

11 National Chicken Council, *Animal Welfare Guidelines and Audit Checklist,* Washington, DC, March 2003, available at www.nationalchickencouncil.com/files/NCCanimalWelfare.pdf. On p. 6 it is stated that "...density shall not exceed 8.5 pounds live weight per square foot." Since the average market weight in 2004 was 5 pounds (see www.nationalchickencouncil.com/statistics/stat_detail.cfm?id=2) this is equivalent to 85 square inches per bird.

12 Compassion in World Farming, www.ciwf.org.uk/campaigns/primary_campaigns/broilersladhousing. html

13 M. O. North and D. D. Bell, *Commercial Chicken Production Manual*, 4th edition (New York: Van Nostrand Reinhold, 1990), p. 456.

14 H. L. Brodie et al, "Structures for Broiler Litter Manure Storage," Fact Sheet 416, Maryland Cooperative Extension www.agnr.umd.edu/users/bioreng/fs416.htm, refer, without any suggestion of criticism, to delaying manure cleanout for 3 years. See also Anon."Animal Waste Management Plans" Delaware Nutrient Management Notes, Delaware Department of Agriculture, vol. 1, no. 7 (July 2000), where the calculations are based on 90 percent of the litter remaining in place for 2 years.

15 C. Berg, "Foot-Pad Dermatitis in Broilers and Turkeys," *Veterinaria* 36 (1998); G. J. Wang, C. Ekstrand, and J. Svedberg, "Wet Litter and Perches as Risk Factors for the Development of Foot Pad Dermatitis in Floor-Housed Hens," *British Poultry Science* 39 (1998): 191–7; C. M. Wathes, "Aerial Emissions from Poultry Production," *World Poultry Science Journal* 54 (1998): 241–51; Kristensen and Wathes, op cit; S. Muirhead, "Ammonia Control Essential to Maintenance of Poultry Health," *Feedstuffs* (April 13, 1992): 11. On blindness caused by ammonia, see also Michael P. Lacy, "Litter Quality and Performance," www.thepoultrysite.com/FeaturedArticle/FATopic.asp?Display=388, and Karen Davis, *Prisoned Chickens, Poisoned Eggs: An Inside Look at the Modern Poultry Industry*, Book Publishing Company, Summertown, TN, 1996, pp. 62–64, 92, 96–98.

16 G. Havenstein, P. Ferket, and M. Qureshi, "Growth, livability, and feed conversion of 1957 versus 2001 broilers when fed representative 1957 and 2001 broiler diets," *Poultry Science* 82 (2003), 1500–1508.

17 S. C. Kestin, T. G. Knowles, A. E. Tinch, and N. G. Gregory, "Prevalence of Leg Weakness in Broiler Chickens and its Relationship with Genotype," *The Veterinary Record* 131 (1992): 190–4.

18 Quoted in *The Guardian*, October 14, 1991; cited in Animals Australia Fact Sheet, "Meat Poultry", http://www.animalsaustralia.org/default2.asp?idL1=1273&idL2=1293.

19 John Webster, *Animal Welfare: A Cool Eye Towards Eden*, Blackwell Science, Oxford, 1995, p. 156.

20 G. T. Tabler and A. M. Mendenhall, "Broiler Nutrition, Feed Intake and Grower Economics," *Avian Advice* 5(4) (Winter 2003), p. 9.

21 J. Mench, "Broiler breeders: feed restriction and welfare, *World's Poultry Science Journal*, vol. 58 (2002), pp. 23–29.

22 I. J. H. Duncan, "The Assessment of Welfare During the Handling and Transport of Broilers," In: J. M. Faure and A. D. Mills (eds.), *Proceedings of the Third European Symposium on Poultry Welfare* (Tours, France: French Branch of the World Poultry Science Association, 1989), pp. 79-91; N. G. Gregory and L. J. Wilkins, "Skeletal Damage and Bone Defects During Catching and Processing," In: *Bone Biology and Skeletal Disorders in Poultry.* C. C. Whitehead, ed. (Carfax Publishing, Abingdom, England, 1992). Cited from A COK Report: Animal Suffering in the Broiler Industry.

23 Freedom of Information Act #94-363, Poultry Slaughtered, Condemned, and Cadavers, 6/30/94; cited in United Poultry Concerns, "Poultry Slaughter: The Need for Legislation", www.upc-online.org/slaughter/slaughter3web.pdf.

24 "Tyson to Probe Chicken-slaughter Methods," Associated Press, May 25, 2005.

25 Signed statement of Tyson employee, Virgil Butler, January 30, 2003. www.kentuckyfriedcruelty.com/virgil/asp. See also Butler's blog, www.cyberactivist.blogspot.com

26 For the video, and other materials, see www.peta.org/feat/moorefield; see also Donald G. McNeil Jr., "KFC Supplier Accused of Animal Cruelty," *New York Times*, July 20, 2004; www.nytimes.com/2004/07/20/business/20chicken.html.

27 John Vidal, *McLibel: Burger Culture on Trial*, Pan Books, London, 1997, p. 311.

28 Chesapeake Bay Foundation, "Fact Sheet: Oysters" www.cbf.org/site/PageServer?pagename=resources_facts_oysters. See also www.chesapeakebay.net/info/american_oyster.cfm.

29 This case study is based on The Chesapeake Bay Foundation, *Manure's Impact on Rivers, Streams and the Chesapeake Bay*, July 28, 2004, www.cbf.org/site/DocServer/

0723manurereport_noembargo_.pdf?docID=2143; and on Peter Goodman, "By-Product: Runoff and Pollution," *The Washington Post*, August 1, 1999, p. A1.

30 Natural Resources and Environmental Protection Cabinet, Department for Environmental Protection, Division of Water, "Statement of Consideration Relating to 401 KAR 5:072—Not Amended After Hearing, June 29, 2000." www.water.ky.gov/NR/rdonlyres/FB5C9A21-1AFA-43CE-B0A7-E27CD967D778/0/REG_SOC.pdf

31 Statement by Sara Shelton and Guy Hardin.

32 Statement by Roger Gamble, Patricia Gamble, Natalie Gamble, Brittany Gamble, Nancy and Roger Grace, Faye Lear, Jean Long, Bernie Miller, Linda McGregor, Ella King.

33 Statement by Linda Moon.

34 *Sierra Club, et al. v. Tyson Foods, Inc., et al*, Case No. 02-CV-073, USDC, Western District of KY; www.sierraclub.org/environmentallaw/lawsuits/viewCase.asp?id=160

35 Sierra Club News Release, "Tyson Chicken Held Accountable for Pollution," January 26, 2005.

36 Alexander Lane, "Egg Farm Neighbors Say System is Broken," *New Jersey Star-Ledger*, October 31, 2004.

37 "Defunct Egg Farm Fined Again," *Marion Online*, February 15, 2005.

38 "Mayor Bill LaFortune's Remarks," House Bill 1879 Press Conference, March 15, 2005, www.cityoftulsa.org/OurCity/Mayor/documents/SupportforAttorneyGeneral-Poultry.pdf

39 Sierra Club, "The Rapsheet on Animal Factories," San Francisco and Washington, DC, 2002.

40 Margaret Stafford, "Tyson Pleads Guilty in Wastewater Case," Associated Press, June 25, 2003.

41 See Mark Kawar, "Tyson, Freddie Mac help workers to buy homes," *Omaha World-Herald*, February 14, 2004, p. 1D, for both the reported turnover figure, and Tyson's refusal to provide the information. Cited from Human Rights Watch, *Blood, Sweat and Fear: Workers' Rights in U.S. Meat and Poultry Plants*, Human Rights Watch, New York, 2004, p. 108n. Tyson also refused to provide workforce turnover figures.

42 Department of Labor News Release, U.S. Newswire, October 13, 1999, cited by Deborah Thompson Eisenberg, "The Feudal Lord in the Kingdom of Big Chicken: Contracting and Worker Exploitation by the Poultry Industry" www.nelp.org/docUploads/eisenberg%2Epdf, p.7n.

43 Kari Lyderson, "Fowl Behavior," *In These Times*, March 19, 2001.

44 Steven Greenhouse, "Unions Finding That Employers Want More Concessions," *The New York Times*, July 11, 2003, p. A12; Cited from Human Rights Watch, *Blood, Sweat and Fear: Workers' Rights in U.S. Meat and Poultry Plants*, Human Rights Watch, New York, 2004, p. 82.

45 Barry Schlachter, "Cooped up: Contract growers hoping the chicken industry offers a steady nest egg may instead be trapped by debt," *Fort Worth Star-Telegram*, May 27, 2005.

46 See, for example, Dennis and Alex Avery, "No More Chicken Run," *Wall Street Journal*, European edition, August 26, 2005.

47 UN News Centre, "UN task forces battle misconceptions of avian flu, mount Indonesian campaign." October 24, 2005, www.un.org/apps/news/story.asp?NewsID=16342&Cr=bird&Cr1=flu.

48 Dennis Bueckert, "Avian flu outbreak raises concerns about factory farms," *Cnews*, April 7, 2004, www.cp.org/english/online/full/agriculture/040407/a040730A.html.

49 Scientific Committee on Animal Health and Animal Welfare, *The Welfare of Chickens Kept for Meat Production (Broilers)*, European Commission, Health and Consumer Protection Directorate General, March 21, 2000.

## CHAPTER 3

1 Ian Duncan, "Welfare Problems of Poultry," in John Benson and Bernard Rollin, eds., *The Well-Being of Farm Animals*, Iowa State Press, Ames, 2004.

2 The videos are available at www.cok.net or on request from Compassion Over Killing.

3 McDonald's comment on Grandin is from their "Global Animal Welfare Progress Report: 2002 Results," www.mcdonalds.com/corp/values/socialrespons/sr_report/progress_report.html

4 Temple Grandin, "Corporations Can Be Agents of Great Improvements in Animal Welfare and Food Safety and the Need for Minimum Decent Standards." A paper presented at the National Institute of Animal Agriculture on April 4, 2001. www.grandin.com/welfare/corporation.agents.html.

5 David Fraser, Joy Mench, Suzanne Millman. "Farm Animals and Their Welfare in 2000," *State of the Animals 2001*, Humane Society Press, 2001, p. 90.

6 United Egg Producers, *Animal Husbandry Guidelines for U.S. Egg Laying Flocks*, 2002, pp. 6–7.

7 Ian Duncan. "The Science of Animal Well-Being." A report from a speech in the Animal Welfare Information Center Newsletter, National Agriculture Library, 1993 (Jan.–March): 4.1, p. 5, as cited in Karen Davis' *Prisoned Chickens, Poisoned Eggs,* Book Publishing Company, 1996, p. 68.

8 "McDonald's & Farming," National Public Radio's "All Things Considered," program aired on April 15, 2002. http://discover.npr.org/features/feature.jhtml?wfId=1141753

9 Mary MacArthur. "Analyst Says Poultry Growers Oblivious to Poor Conditions," *Western Producer*, Dec. 12, 2002.

10 "U.S. Egg Producers to Phase Out Feed Withdrawal," *Food Production Daily*, May 27, 2005; www.foodproductiondaily.com/news/printNewsBis.asp?id=60285

11 Alexei Barrionuevo, "Egg Producers Relent on Industry Seal," *New York Times*, October 4, 2005.

## CHAPTER 4

1 Christopher G. Davis and Biing-Hwan Lin, "Factors Affecting US Pork Consumption," Economic Research Service, U.S. Department of Agriculture, Outlook Report No. (LDPM13001), May 2005, www.ers.usda.gov/publications/LDP/may05/ldpm13001/ldpm13001.pdf

2 http://statistics.defra.gov.uk/esg/publications/nfs/1999/sections5.pdf.

3 *Corporate Fact Sheet*; Overview, Kraft Foods. http://kraft.com/profile/factsheet.html

4 Renee Zahery, telephone message, February 1, 2005.

5 Ronald L. Plain, "Trends in U.S. Swine Industry," paper for U.S. Meat Export Federation Pork Conference, Taipei, Taiwan, September 24, 1997. www.ssu.missouri.edu/faculty/RPlain/papers/swine.htm; T. Stout and G. Packer, "National Trends Reflected in Changing Ohio Swine Industry," Ohio State University Extension Research Bulletin, Special Circular 156, Agricultural Economics Department, (n. d.) http://ohioline.osu.edu/sc156/sc156_48.html

6 U.S. Department of Agriculture, National Agricultural Statistics Service, Livestock Slaughter, 2004 Summary, March, 2005. http://usda.mannlib.cornell.edu/reports/nassr/livestock/pls-bban/lsan0305.pdf. The decline in pig farm numbers averages about seven percent per year.

7 Environmental Defense, "Factory Hog Farming: The Big Picture," November 2000, www.environmentaldefense.org/documents/2563_FactoryHogFarmingBigPicture.pdf.

8 Lynn Bonner, "Critics Say State Must Do More to Protect Rivers," *Raleigh News & Observer*, 17 August 1995; Minority Staff, U.S. Senate Committee on Agriculture, Nutrition and Forestry, "Animal Water Pollution in America: An Emerging National Problem, " 105th Congress, 1st session, December 1997, p. 3. We owe these references to Carolyn Johnsen, *Raising a Stink: The Struggle Over Factory Hog Farms in Nebraska*, University of Nebraska Press, Lincoln, 2003, pp. 14–15.

9 Carolyn Johnsen, *Raising a Stink: The Struggle Over Factory Hog Farms in Nebraska*, University of Nebraska Press, Lincoln, 2003, pp. 21–26.

10 Paul Hammel, "Turning Hog Odors into Tax Deductions," *Omaha World-Herald*, March 5, 2002, cited in Carolyn Johnsen *Raising a Stink: The Struggle Over Factory Hog Farms in Nebraska*, University of Nebraska Press, Lincoln, 2003, p. 138.

11 American Public Health Association, "Precautionary Moratorium on New Concentrated Animal Feed Operations," *2003 Policy Statements*, pp. 12-14, www.apha.org/legislative/policy/2003/2003-007.pdf.

12 Ross Clark, "If only pigs could talk," *Sunday Telegraph* (London) March 23, 1997; Roger Highfield, "Computer Skills Show Just How Smart Pigs Are," *Ottawa Citizen* May 29, 1997. (originally published in the *Daily Telegraph*, London.)

13 David Wolfson, *Beyond the Law: Agribusiness and the Systemic Abuse of Animals Raised for Food or Food Production,* Watkins Glen, NY: Farm Sanctuary, Inc., 1999. See also "COK Talks with David Wolfson, Esq." www.cok.net/abol/16/04.php

14 On the number of pigs kept indoors, see National Animal Health Monitoring System, Animal and Plant Health Inspection Service, U.S. Department of Agriculture, *Swine 2000*, Part I: Reference of Swine Health and Management in the United States, 2000, Washington, DC, 2001, p. 26. Very few total confinement systems in the U.S. use straw or any other form of bedding. www.aphis.usda.gov/vs/ceah/ncahs/nahms/swine/swine2000/Swine2kPt1.pdf.

15 On the number of sows in crates in the ten biggest producers, see U.S. Department of Agriculture, Agricultural Research Service, Livestock Issues Research, "Research Project: The Emerging Issue of Sow Housing," 2004 Annual Report. The overall estimate is from Glenn Grimes, professor emeritus of agricultural economics, University of Missouri, interview with Jim Mason, July 5, 2005.

16 See Scientific Veterinary Committee, Animal Welfare Section, The Welfare of Intensively Kept Pigs, 1997, and Clare Druce and Philip Lymbery, "Outlawed in Europe," in Peter Singer, ed., *In Defense of Animals: The Second Wave*, Blackwell, Oxford, 2005.

17 In the European Union, 242 million pigs were slaughtered in 2005, compared to 103 million in the U.S. See "EU output data revised," *Pig International Electronic Newsletter*, June 23, 2005, based on Eurostat information, www.wattnet.com/newsletters/Pig/htm/jun05pigenews.htm.

18 Per Jensen, "Observations on the Maternal Behaviour of Free-Ranging Domestic Pigs" *Applied Animal Behaviour Science*, vol.16 (1986) pp. 131–42.

19 Governor's Office of Drug Control Policy, Iowa, "Iowa METH Facts," February 23, 2005, www.state.ia.us/government/odcp/docs/Meth_Other_Drug_Facts_Feb23.pdf.

20 Bernard Rollin first reported on this case in his column in *Canadian Veterinary Journal*, 32:10 (October 1991), p.584; the column is reprinted in Bernard Rollin, *Introduction to Veterinary Medical Ethics: Theory and Cases*, Blackwell, Oxford, 1999.

21 Jonathan Leake, "The Secret Lives of Moody Cows," *Sunday Times*, February 27, 2005.

22 Peter Lovenheim, *Portrait of a Burger as a Young Calf*, Three Rivers Press, New York, 2002. We are grateful to Peter Lovenheim for checking our text and clarifying some issues.

23 John Peck, "Dairy Farmer Workers Fight for Their Rights in Oregon," *Z Magazine Online*, vol. 17, no. 12 (December 2004), http://zmagsite.zmag.org/Dec2004/peckpr1204.html; www.braums.com/FAQ.asp#9

24 Eddy LaDue, Brent Gloy, and Charles Cuykendall, "Future Structure of the Dairy Industry: Historical Trends, Projections and Issues," Cornell University, Ithaca, NY, June 2003, http://aem.cornell.edu/research/researchpdf/rb0301.pdf, p.iii.

25 To be precise, the increase is from 665 gallons a year in 1950 to 2,365 gallons per year in 2004, an increase of 355 percent. See Erik Marcus, *Meat Market*, Brio Press, Ithaca, NY, 2005, pp. 10-11, drawing on figures from USDA National Agricultural Statistical Services, and updated from http://usda.mannlib.cornell.edu/reports/nassr/dairy/pmp-bb/2005/mkpr0105.txt.

26 USDA, National Animal Health Monitoring System, *Dairy 2002*, Part I: Reference of Dairy Health and Management in the United States, p. 54, www.aphis.usda.gov/vs/ceah/ncahs/nahms/dairy/dairy02/Dairy02Pt1.pdf.

27 Peter Lovenheim, *Portrait of a Burger as a Young Calf*, Three Rivers Press, New York, 2002, p. 87.

28 Peter Lovenheim, *Portrait of a Burger as a Young Calf*, Three Rivers Press, New York, 2002, p. 16.

29 Oliver Sacks, *An Anthropologist on Mars*, Knopf, New York, 1995, p. 267.

30 Quoted from People for the Ethical Treatment of Animals, "Cows Grieve," www.goveg.com/f-hiddenlivescows_giants.asp.

31 Jon Bonné, "Can Animals You Eat Be Treated Humanely?" MSNBC News, June 28, 2004, http://msnbc.msn.com/id/5271434/

32 Peter Lovenheim, *Portrait of a Burger as a Young Calf*, Three Rivers Press, New York, 2002, pp. 112–113.

33 Miguel Bustillo, "In San Joaquin Valley, Cows Pass Cars as Polluters," *Los Angeles Times*, August 2, 2005.

34 Michael Pollan, "Power Steer," *The New York Times Sunday Magazine*, March 31, 2002.

35 "Researchers, McDonald's Say U.S. Govt BSE Defense Not Working," *Cattlenetwork.com,* January 4, 2006, www.cattlenetwork.com/content.asp?contentid=16082.

36 Chris Clayton, "More than 1250 Nebraska Cattle Died in Heat Wave," *Omaha World-Herald*, July 27, 2005.

37 F. M. Mitlöhner, et al, "Effects of shade on heat-stressed heifers housed under feedlot conditions," *Burnett Center Internet Progress Report*, no. 11, February 2001; www.depts.ttu.edu/liru_afs/pdf/bc11.pdf; see also F. M. Mitlöhner, et al, "Shade effects on performance, carcass traits, physiology, and behavior of heat-stressed feedlot heifers," *Journal of Animal Science*, vol. 80 (2002) pp. 2043–2050, http://jas.fass.org/cgi/content/full/80/8/2043.

38 A. M. Soto et al, "Androgenic and estrogenic activity in cattle feedlot effluent receiving water bodies of eastern Nebraska, USA." *Environmental Health Perspectives*. 112 (2004), pp. 346–352; E. F. Orlando et al, "Endocrine disrupting effects of cattle feedlot effluent on an aquatic sentinel species, the fathead minnow." *Environmental Health Perspectives*. 112 (2004), pp. 353–358; Janet Raloff, "Hormones: Here's the Beef," *Science News*, Vol 161, (Jan. 5, 2002), p. 10. www.sciencenews.org/articles/20020105/bob13.asp

39 Carolyn Johnsen, *Raising a Stink: The Struggle Over Factory Hog Farms in Nebraska*, University of Nebraska Press, Lincoln, 2003, p. 24.

40 "EPA says it will inspect Idaho feedlots," *Cow-Calf Weekly* (BEEF), August 5, 2005.

41 U.S. Environmental Protection Agency, Region 5, *Results of an Informal Investigation of The National Pollutant Discharge Elimination System Program for Concentrated Animal Feeding Operations in the State of Michigan*, Interim Report, July 24, 2002; we owe the reference to Tony Dutzik, The State of Environmental Enforcement, CoPIRG Foundation, Denver, 2002, www.environmentcolorado.org/reports/envenfco10_02.pdf, which discusses the problem of lack of state environmental enforcement.

42 Nebraska Department of Environmental Quality, Water Quality Division, 2002 Nebraska Water Quality Report, Lincoln, 2002, cited by Carolyn Johnsen, *Raising a Stink: The Struggle Over Factory Hog Farms in Nebraska*, University of Nebraska Press, Lincoln, 2003, p. 138.

43 Carolyn Johnsen, *Raising a Stink: The Struggle Over Factory Hog Farms in Nebraska*, University of Nebraska Press, Lincoln, 2003, p. 122.

44 U.S. General Accounting Office, *Humane Methods of Slaughter Act*, January 2004, www.gao.gov/new.items/d04247.pdf

45 "AgriProcessors," video available on the Web site of People for the Ethical Treatment of Animals, www.petatv.com/inv.html

46 Sholem Rubashkin, "Response." Shmais News Service, no date, www.shmais.com/jnewsdetail.cfm?ID=148; Department of Public Relations, Orthodox Union, "Orthodox Union Releases Industry Animal Welfare Audit of Agriprocessors," March 7, 2005, www.ou.org/oupr/2005/agri65.htm

## CHAPTER 5

1 Quoted by Milton Moskovitz, "That's the Spirit," *Mother Jones*, July/August 1997, www.motherjones.com/news/feature/1997/07/moskowitz.html

2 McDonald's Corporation, "Interview with Dr. Temple Grandin," www.mcdonalds.com/corp/values/socialrespons/sr_report/progress_report/grandin_interview.html

3 McDonald's Animal Welfare Program, Australia, www.rmhc.org/corp/values/socialrespons/sr_report/progress_report/australia.html

4 Associated Press, August 23, 2000; *Los Angeles Times*, September 7, 2000.

5 Food Marketing Institute, www.fmi.org/media/mediatext.cfm?id=522

6 McDonald's Corporation, "Interview with Dr. Temple Grandin," www.mcdonalds.com/corp/values/socialrespons/sr_report/progress_report/grandin_interview.html.

7 David Fraser, Joy Mench, Suzanne Millman. "Farm Animals and Their Welfare in 2000," *State of the Animals 2001*, Humane Society Press, 2001, p. 90.

8 Kim Severson, "Humane Handling Taking Hold on Farms," *San Francisco Chronicle*, September 7, 2003, http://sfgate.com/cgi-bin/article.cgi?file=/chronicle/archive/2003/09/07/MN165897.DTL; interview with Diane Halverson, July 2005.

9 For details see www.ansc.purdue.edu/CAWB/.

10 *McDonald's Global Policy on Antibiotic Use in Food Animals*, June 3, 2003, www.mcdonalds.com/corp/values/socialrespons/market/antibiotics/global_policy.html.

11 Associated Press, "Citing the Human Threat, U.S. Bans a Poultry Drug," *New York Times*, July 29, 2005.

12 See www.mcdonalds.com/corp/values/socialrespons/sr_report.html

13 Emory Heart Center, "What Would 100 Billion McVeggie Burgers Mean? Healthier Customers, Study Says," April 18, 2005. www.emoryhealthcare.org/news_events/press_room/ehc_news/McVeggie_Burgers.html.

14 Jennifer Waters, "Wal-Mart Grocery Share Seen Doubling," *CBS MarketWatch*, June 24, 2004, cited from RetailWire Discussions, www.retailwire.com/Discussions/Sngl_Discussion.cfm/9953.

15 Michael Barbaro, "A New Weapon for Wal-Mart: A War Room," *New York Times*, November 1, 2005.

16 John Dicker, *The United States of Wal-Mart*, Tarcher/Penguin, New York, 2005, p. 122.

17 Pankaj Ghemawat and Ken Mark, "The Price is Right," *New York Times*, August 3, 2005.

18 John Dicker, *The United States of Wal-Mart*, Tarcher/Penguin, New York, 2005, p. 86.

19 Michael Barbaro, "Wal-Mart to Expand Health Plan for Workers," *New York Times*, October 24, 2005; Reed Abelson, "Wal-Mart's Health Care Struggle is Corporate America's Too," *New York Times*, October 29, 2005.

20 "Wal-Mart Sets the Record Straight" www.walmartfacts.com/newsdesk/article.aspx?id=1091

21 Charles Fishman, "The Wal-Mart you don't know," *Fast Company*, December 2003, www.fastcompany.com/magazine/77/walmart.html; the quote from Gib Carey is also from this article.

22 Wal-Mart Stores, Inc., "Standards for Suppliers," www.walmartstores.com/Files/SupplierStandards-June2005.pdf

23 Food Marketing Institute, "Status FMI-NCCR Animal Welfare Guidelines, Updated May 2005," www.fmi.org/animal_welfare/guideline_status_chart_May_2005.pdf.

24 John Dicker, *The United States of Wal-Mart*, Tarcher/Penguin, New York, 2005, p. 213.

## CHAPTER 6

1 Roger Scruton, "The Conscientious Carnivore," in Steve Sapontzis, ed., *Food for Thought: The Debate over Eating Meat*, Prometheus, Amherst, NY, 2004.

## CHAPTER 7

1 Tim Holmes, interview, May 2, 2005.

2 A typical crate is two feet wide and seven feet long, or fourteen square feet. Of course crates do not take up the entire floor space inside a factory gestation building. Typically, they are arranged side-to-side in two rows down the length of the building with a four- or five-foot-wide aisle between the rows and narrower walkways around the walls to permit power washing and maintenance. The building is about 30' by 10' and holds about 100 sows. One could put three such buildings, then, on the space that the Holmes family allows one of their gestating sows.

3 A. Stolba and D. G. Wood-Gush, "The Behaviour of Pigs in a Semi-Natural Environment," *Animal Production*, vol. 48 (1989) pp. 419–425.

4 Diane Halverson, interview, May 2, 2005.

## CHAPTER 8

1 Jia-Rui Chong, "Vet in row after hens 'chipped' to death," *Los Angeles Times*, November 23, 2003; "Abuse charges filed against Moark egg company." *News Tribune* (Jefferson City, MO) July 30, 2005; "Moark must pay $100,000," www.hsus.org/farm_animals/farm_animal_news/Moark_Settles_Case.html. (October 25, 2005)

2 Interview, Nick Levendoski and David Bruce, August 2005.

3 The Soil Association, *Soil Association Chickens Rule the Roost*, http://www.soilassociation.org/web/sa/saweb.nsf/a71fa2b6e2b6d3e980256a6c004542b4/674c00f8ade6722c80257031004b1c54!OpenDocument&Highlight=2,chickens

4 See Helen Thomas, "The Free Range Fiddle," Background Briefing, Radio National, Australian Broadcasting Corporation, June 26, 2005, www.abc.net.au/rn/talks/bbing/stories/s1397934.htm.

## CHAPTER 9

1 Marine Stewardship Council, "Fish Facts" http://eng.msc.org/html/content_528.htm. The 17 billion estimate is from www.fishinghurts.com/fishing101.asp, calculated by dividing the total weight of seafood consumed by an estimated average weight per creature.

2 "Challenge to Fishing: Keep Unwanted Species Out of Its Huge Nets," Otto Pohl, *The New York Times*, July 29, 2003 www.nytimes.com/2003/07/29/science/29BYCA.html.

3 Garrett Hardin, "The Tragedy of the Commons," *Science*, 162 (1968) pp. 1243–48.

4 "A Run on the Banks: How 'Factory Fishing' Decimated Newfoundland Cod," Colin Woodward, *E Magazine*, March/April, 2001. www.emagazine.com/view/?507.

5 Information from Tim Fitzgerald, Environmental Defense Trust, Oceans Program, 257 Park Avenue South, New York, NY 10010, and James Ianelli, Alaska Fish. Science Center, Seattle, March/April 2005. See also Pacific Rim Fisheries Program, Institute of the North, Alaska Pacific University, http://prfisheries.alaskapacific.edu/PRF_Statistics/usa/usa_fish_species_federal.htm.

6 Marine Stewardship Council http://eng.msc.org/html/content_1188.htm.

7 The statement was on the Web site of Angelina's of Maryland, www.crabcake.com, in 2005 but has subsequently been removed from the site.

8 Email message from Nancy Rosenberg to Jim Mason, Thursday, March 10, 2005.

9 www.blue-crab.org/spawning.htm

10 Jose A. Ingles. Biological and Fisheries Assessment of the Blue Crab Resources of the Northeastern Guimaras Strait. A final report submitted to WWF-Philippines, 2000; K .L.Jayme, G. Romero and J. A. Ingles, 2003, "Community-based certification of the blue crab fishery of the northeastern Guimaras Strait, Negros Occidental, Philippines: lessons learned, prospects and directions." Paper presented at the Second International Tropical Marine Ecosystems Management Symposium (ITMEMS 2), Manilla, Philippines. March 24–27, 2003. Theme 11. www.reefbase.org/References/ref_literature_detail.asp?refID=14923; and Katrina Jayme, email communication, April 2005.

11 P. Redmaynem "Blue crab: Asian import," *IntraFish*, February 2004, p. 20 www.intrafish.com/pdf/download/2a71bcdc0cc482441f99-7ea72cea1f35/2004/2/20.pdf

12 Email message from Brenda Davis to Jim Mason, Wednesday March 19, 2005. Dr. Davis is a Fisheries Biologist with Maryland's Department of Natural Resources, Stevensville, MD. See also, "King Crab," by Jon Goldstein, *Baltimore Sun*, Business, July 2, 2001.

13 John Ryan, "Feedlots of the Sea," *WorldWatch Magazine*, September/October 2003, www.worldwatch.org/pubs/mag/2003/165; WorldWatch Institute, "Factory-Fish Farming," *WorldWatch Magazine*, September/October 2003, www.worldwatch.org/pubs/mag/2003/165/mos/.

14 Marian Burros, "Stores Say Wild Salmon, but Tests Say Farm Bred," *The New York Times*, April 10, 2005.

15 WorldWatch Institute, "Factory-Fish Farming," *WorldWatch Magazine*, September/October 2003, www.worldwatch.org/pubs/mag/2003/165/mos/, using data from the U.S. Department of Agriculture and the Fisheries Department, Food and Agriculture Organization of the United Nations.

16 Philip Lymbery, "In Too Deep—The Welfare of Intensively Farmed Fish," Compassion in World Farming Trust, Petersfield, Hampshire, 2002. p. 3.

17 Juliet Eilperin, "Fish Farming's Bounty Isn't Without Barbs," *Washington Post*, January 24, 2005, www.washingtonpost.com/wp-dyn/articles/A31159-2005Jan23.html.

18 Daniel Pauly, et al, "Towards sustainability in world fisheries," *Nature*, vol. 418, pp. 689–695.

19 Rees and Tydeman are quoted in John Ryan, "Feedlots of the Sea," *WorldWatch Magazine*, September/October 2003, pp. 22-29, www.worldwatch.org/pubs/mag/2003/165.

20 Kenneth Weiss, "Bush Proposal Seeks to Cultivate Fish Farming in Federal Waters, *Los Angeles Times*, June 8, 2005.

21 Juliet Eilperin, "Fish Farming's Bounty Isn't Without Barbs," *Washington Post*, January 24, 2005, www.washingtonpost.com/wp-dyn/articles/A31159-2005Jan23.html.

22 John Ryan, "Feedlots of the Sea," *WorldWatch Magazine*, September/October 2003, pp. 22–29, www.worldwatch.org/pubs/mag/2003/165; for a fuller account, see "Sea cage fish farming: an evaluation of environmental and public health aspects (the five fundamental flaws of sea cage fish farming), "a paper presented by Don Staniford at the European Parliament's Committee on

Fisheries public hearing on Aquaculture in the European Union: Present Situation and Future Prospects," October 1, 2002: www.watershed-watch.org/ww/publications/sf/Staniford_Flaws_SeaCage.PDF.

23 "Farm Sea Lice Plague Wild Salmon," *BBC News*, March 29, 2005, http://news.bbc.co.uk/2/hi/science/nature/4391711.stm.

24 www.oceansalive.org/eat.cfm?subnav=fishpage&fish=85

25 www.mbayaq.org/cr/SeafoodWatch/websfw_factsheet.aspx?fid=27

26 Seafood Watch *Seafood Report, Shrimp* vol. III, *Farm Raised Shrimp, World Overview*, prepared by Alice Cascorbi, Monterey Bay Aquarium, 2004, p. iv, based on information from Jason Clay, Senior Fellow, World Wildlife Fund-US. www.mbayaq.org/cr/cr_seafoodwatch/content/media/MBA_SeafoodWatch_FarmedShrimpReport.pdf

27 Ibid.

28 Ibid.

29 Quoted in Seafood Watch *Seafood Report, Shrimp* vol III, *Wild-Caught Warmwater Shrimp*, prepared by Alice Cascorbi, Monterey Bay Aquarium, 2004, p. 17.

30 U.S. National Marine Fisheries, Report to Congress, 2003: Status of Fisheries of the United States for 2002. Published April 2003. National Marine Fisheries Service, Silver Spring, MD, cited in *Seafood Watch Seafood Report, Shrimp* vol. III, Wild-Caught Warmwater Shrimp, prepared by Alice Cascorbi, Monterey Bay Aquarium, 2004, pp. 15, 17.

31 Barry Evans, "Making the Best Even Better," Australian Government, Fisheries Research and Development Corporation, *R&D* news, 13:3 www.frdc.com.au/pub/news/133.01.php.

32 J. B. Robbins, M. J. Campbell and J. G. McGilvray, "Reducing Praw-trawl Bycatch in Australia: An Overview and an Example from Queensland," *Marine Fisheries Review*, 61:3 (1999), p. 46.

33 *Good Stuff*, WorldWatch Institute, Washington, DC, 2004, sec. 26, www.worldwatch.org/pubs/goodstuff/shrimp/.

34 *Seafood Watch Seafood Report, Shrimp* vol. III, Farm Raised Shrimp, World Overview, prepared by Alice Cascorbi, Monterey Bay Aquarium, 2004, p. 11, citing Thor Lassen, "Mangrove conservation and shrimp aquaculture."Presentation to World Aquaculture Society, March 1-5, 2004, Honolulu, Hawaii. www.mbayaq.org/cr/cr_seafoodwatch/content/media/MBA_SeafoodWatch_FarmedShrimpReport.pdf.

35 Acción Ecológica Declaración De Majagual. February 2003. www.accionecologica.org/sobeali3.htm; Taylor, N. "Hungry for change: Protected area and Ramsar Site 1000, La Barberie, has been destroyed by the shrimp company El Faro." *Honduras This Week* Online. April 14, 2003; cited in "Shrimp's Passport: How International Trade Agencies Monitor America's Favorite Seafood," *Public Citizen*, Washington, DC, 2005, p. 12, www.citizen.org/cmep/foodsafety/shrimp/articles.cfm?ID=12971

36 Indian Supreme Court, Petitioner: S. Jagannath Vs. Respondent: Union Of India & Ors. Date of Judgment. November 12, 1996., cited in "Shrimp's Passport: How International Trade Agencies Monitor America's Favorite Seafood," *Public Citizen*, Washington, DC, 2005, p. 18, www.shrimpactivist.org.

37 Seafood Watch *Seafood Report, Shrimp* vol. III, *Farm Raised Shrimp, World Overview*, prepared by Alice Cascorbi, Monterey Bay Aquarium, 2004, p.11, citing A Wistrand, "Shrimp Cultivation Puts Environment in Danger," published by Nijera Kori, a Bangladesh NGO. www.mbayaq.org/cr/cr_seafoodwatch/content/media/MBA_SeafoodWatch_FarmedShrimpReport.pdf.

38 Darry Jory, "Shrimp Farming in Venezuela: A Case Study," presented to the World Aquaculture Society Conference, Brazil, May 20, 2003 www.iiap.org.pe/publicaciones/CDs/CONFERENCIAS_WAS/WAS_BRASIL/Brazil/.

39 www.mbayaq.org/cr/seafoodwatch.asp

40 Alex Kirby, "Prawn fishing 'plundering seas'", BBC News, February 19, http://news.bbc.co.uk/2/hi/science/nature/2776359.stm.

41 S. D. Sedgwick, *Salmon Farming Handbook,* Fishing News Books, Surrey, 1988, quoted in Philip Lymbery, "In Too Deep—The Welfare of Intensively Farmed Fish." Compassion in World Farming Trust, Petersfield, Hampshire, 2002, p. 17.

42 Philip Lymbery, "In Too Deep—The Welfare of Intensively Farmed Fish." Compassion in World Farming Trust, Petersfield, Hampshire, 2002.

43 Dawnwatch, "CBS Hit Series "Judging Amy" Looks at Animal Cruelty," April 12, 2005, www.dawnwatch.com/4-05_Animal_Media_Alerts#JUDGING _AMY

44 James D. Rose, "The Neurobehavioral Nature of Fishes and the Question of Awareness and Pain," *Reviews in Fisheries Science.* 10 No. 1, (2002), pp. 1–38, uwadmnweb.uwyo.edu/Zoology/faculty/Rose/pain.pdf.

45 Lynne Sneddon, V. A. Braithwaite, and M . J. Gentle, "Do fish have nociceptors? Evidence for the evolution of a vertebrate sensory system." *Proceedings of the Royal Society* vol. 270, No. 1520 (2003), pp. 1115–1121. See also "Trout Trauma Puts Anglers on the Hook?" *Science News,* April 30, 2003, www.royalsoc.ac.uk/news.asp?year=&id=1697 and Sanjida O'Connell, "Does she have feelings too?" *Daily Telegraph,* March 3, 2005.

46 Culum Brown, "Not just a pretty face," *New Scientist*, vol. 182, no. 2451, June 12, 2004, p. 42. See also K. P. Chandroo, I. J. H. Duncan, and R .D. Moccia, "Can fish suffer? Perspectives on sentience, pain, fear and stress," *Applied Animal Behaviour Science,* vol. 86 (2004) pp. 225–250.

## CHAPTER 10

1 Rich Pirog, et al, *Food, Fuel and, Freeways,* Leopold Center for Sustainable Agriculture, University of Iowa, Ames, Iowa, 2001. Available at www.leopold.iastate.edu/pubs/staff/ppp/food_mil.pdf.

2 The description of the Wal-Mart distribution center at Bentonville comes from Associated Press, "Vice President Cheney Visits Wal-Mart's Hometown," but for an account of Wal-Mart's distribution system, with similar distances covered, see Brian Halweil, *Eat Here,* Norton, New York, 2004, p. 7.

3 Erik Millstone and Tim Lang, *The Atlas of Food,* Earthscan, London, 1963, p. 60.

4 Rich Pirog, et al, *Food, Fuel and, Freeways,* Leopold Center for Sustainable Agriculture, University of Iowa, Ames, Iowa, 2001. Available at www.leopold.iastate.edu/pubs/staff/ppp/food_mil.pdf.

5 Andy Jones, *Eating Oil,* Sustain & Elm Farm Research Centre, London, 2001, Case Study 1. www.sustainweb.org/chain_fm_eat.asp

6 Nick Marathon, Tamera VanWechel, and Kimberly Vachal. *Transportation of U.S. Grains: A Modal Share Analysis, 1978-95.* U.S. Department of Agriculture, Washington, DC, 2004. www.ams.usda.gov/tmd/TSB/Modal_Share.pdf; FAO, *FAOSTAT: Commodity Balances*, 2004. http://apps.fao.org/faostat/collections?version=ext&hasbulk=0&subset=agriculture.

7 K. Klindworth, *Agricultural Transportation Challenges for the 21st Century: A Framework for Discussion.* U.S. Department of Agriculture, AMS Transportation and Marketing Programs, 1999. Cited in M. Hora and J. Tick, *From Farm to Table: Making the Connection in the Mid-Atlantic Food System*: Capital Area Food Bank, Washington, DC, 2001.

8 The Council on the Environment of the City of New York, "Greenmarket Farmers Market," www.cenyc.org/HTMLGM/maingm.htm.

9 www.ams.usda.gov/farmersmarkets; www.foodroutes.org

10 Robert Summer, et al, "The Behavioral Ecology of Supermarkets and Farmers Markets," *Journal of Behavioral Psychology*, vol. 1, March 1981, pp. 13–19, and more recent unpublished studies by Robert Sommers, cited by Brian Halweil, *Eat Here*, Norton, New York, 2004, p. 10.

11 *Farmers' Markets: a business survey,* National Farmers' Union, London, September 2002.

12 Brian Halweil, *Eat Here*, p.165; Richard Evanoff, "A look inside Japan's Seikatsu Club Consumers' Cooperative," *Social Anarchism*, No. 26, (1998), http://library.nothingness.org/articles/all/en/display/247.

13 Jeanette Lee, "Colleges Buying More Food From Farmers," Associated Press, January 27, 2005; taken from *Mercury News*, San Jose, CA Yale Sustainable Food Project, www.yale.edu/sustainablefood, and Alison Leigh Cowan, "A Dining Hall Where the Students Try to Sneak In," *New York Times*, May 10, 2005.

14 The indented passages come from www.FoodRoutes.org, June 2005.

15 U.S. Department of Agriculture, National Agricultural Statistics Service, *Trends in U.S. Agriculture, Labor Force and Farm Labor*, 1900-1990, www.usda.gov/nass/pubs/trends/farmpopulation.htm; USDA/NASS, *2002 Census of Agriculture*, Vol. 1, Table 1, www.nass.usda.gov/census/census02/volume1/us/st99_1_001_001.pdf 60; Bureau of Justice Statistics, Press Release, "Prison Population Approaches 1.5 million", November 7, 2004, www.ojp.usdoj.gov/bjs/pub/press/p03pr.htm. (I have drawn on Brian Halweil, *Eat Here*. p. 199, n.6 for this comparison.)

16 Bill Vorley, *Food Inc.*, International Institute for Environment and Development, London, 2003, p. 9.

17 Timothy Egan, "Amid Dying Towns of Rural Plains, One Makes a Stand," *The New York Times*, December 1, 2003.

18 Jon Bailey and Kim Preston, *Swept Away: Chronic Hardship and Fresh Promise on the Great Plains*, Center for Rural Affairs, Walthill, Nebraska, 2003, Pt I.

19 Rich Pirog, et al, *Food, Fuel and, Freeways*, Leopold Center for Sustainable Agriculture, University of Iowa, Ames, Iowa, 2001. Available at www.leopold.iastate.edu/pubs/staff/ppp/food_mil.pdf.

20 Verlyn Klinkenborg, "Keeping Iowa's Young Folks at Home After They've Seen Minnesota," *The New York Times*, February 8, 2005.

21 U.S. Department of Agriculture, *Agriculture Fact Book 98*, www.usda.gov/news/pubs/fbook98/chart1.htm, fig 1-8; *Economic Research Service, USDA*, "Food Marketing and Price Spreads: USDA Marketing Bill," www.ers.usda.gov/Briefing/FoodPriceSpreads/bill/table1.htm.

22 Michael Rosmann, AgriWellness Inc., and Paul Gunderson, National Farm Medicine Center, in discussion with Brian Halweil; Karen Pylka and Paul Gunderson, "An Epidemiologic Study of Suicide Among Farmers and Its Clinical Implications," *Marshfield Clinical Bulletin*, vol. 26, 1992, pp. 31–58. I owe these references to Brian Halweil, *Eat Here*, pp. 69–70.

23 Intergovernmental Panel on Climate Change, *Third Assessment Report: Summary for Policymakers: The Science of Climate Change*. IPCC Working Group I, p. 10. Available at: www.ipcc.ch. On the hottest years, see Traci Watson, "2004 is 4th hottest year for world since 1861, U.N. report says," *USA Today*, December 15, 2004, www.usatoday.com/weather/news/2004-12-15-hot-year_x.htm

24 Richard Posner, *Catastrophe: Risk and Response*, Oxford University Press, New York, 2004.

25 For further discussion of climate change as an ethical issue, see Peter Singer, *One World*, Yale University Press, 2002, chapter 2.

26 John Hendrickson, "Energy use in the U.S. Food System: A summary of existing research and analysis." *Sustainable Farming* (Ste. Anne de Bellevue, Quebec), vol. 7, no. 4. Fall 1997.

27 G. Schueller, "Eat Local." *Discover*. 22(5), 2001.

28 John Hendrickson, "Energy use in the U.S. Food System: A summary of existing research and analysis." *Sustainable Farming* (Ste. Anne de Bellevue, Quebec), vol. 7, no. 4. Fall 1997.

29 For the number of BTUs it takes to move freight by road, see U.S. Congress, Office of Technology Assessment, *Saving Energy in U.S. Transportation*, OTA-ETI-589 (Washington, DC: U.S. Government Printing Office, July 1994), p. 44, http://govinfo.library.unt.edu/ota/Ota_1/DATA/1994/9432.PDF.

30 Alison Smith, et al, *The Validity of Food Miles as an Indicator of Sustainable Development*, ED50254, Issue 7, July 2005, p.67; A. Carlsson , *Greenhouse Gas Emissions in the Life-Cycle of Carrots and Tomatoes: methods, data and results from a study of the types and amounts of carrots and tomatoes consumed in Sweden*, IMES/EESS Report no. 24, Department of Environmental and Energy Systems Studies, Lund University, Sweden, March 1997; cited in Tara Garnett, *Wise Moves*, Transport 2000, pp. 76, 82–4.

31 Intergovernmental Panel on Climate Change, *Aviation and the Global Atmosphere,* Cambridge University Press, 1999; J. Whitelegg and N. Williams, *The Plane Truth: aviation and the environment*, Transport 2000 and Ashden Trust, London, 2001; we owe these references to Tara Garnett, *Wise Moves*, Transport 2000, p. 23.

32 J. Pretty and A. Ball, "Agricultural Influences on Carbon Emissions and Sequestration: A Review of Evidence and the Emerging Trading Options," Centre for Environment and Society Occasional Paper 2001-03, University of Essex, 2001.

33 Andy Jones, *Eating Oil*, Sustain & Elm Farm Research Centre, London, 2001, Case Study 2. www.sustainweb.org/chain_fm_eat.asp.

34 Alison Smith, et al, *The Validity of Food Miles as an Indicator of Sustainable Development*, ED50254, Issue 7, July 2005, p. 74.

35 Email from Carlo Petrini to Brian Halweil, cited in Brian Halweil, *Eat Here,* p. 161.

## CHAPTER 11

1 Diana Friedman, "The Del Cabo project; a Mexican collective exports organic produce to the U.S.A.," *Whole Earth Review,* Spring 1989. www.findarticles.com/p/articles/mi_m1510/is_n62/ai_7422469; Don Lotter, "The Del Cabo Cooperative of Southern Baja keeps 300 farm families busy growing organic crops for export," *New Farm,* July 20, 2004, www.newfarm.org/international/pan-am_don/july04/.

2 *United Nations Human Development Report, 2005,* p. 24. http://hdr.undp.org/reports/global/2005/pdf/HDR05_chapter_1.pdf

3 United Nations Development Programme, *Human Development Report* 2000 (Oxford University Press, New York, 2000), p. 30; *Human Development Report* 2001 (Oxford University Press, New York, 2001), pp. 9–12, p. 22; and *World Bank, World Development Report* 2000/2001, Overview, p. 3, www.worldbank.org/poverty/wdrpoverty/report/overview.pdf, for the other figures. The *Human Development Reports* are available at http://hdr.undp.org.

4 Nomaan Majid, "Reaching Millennium Goals: How well does agricultural productivity growth reduce poverty?" Employment Strategy Papers, International Labor Organization, 2004, http://www.ilo.org/public/english/employment/strat/download/esp12.pdf.

5 M. Ataman Aksoy and John Beghin, eds., Global Agricultural Trade and Developing Countries, World Bank, Washington, DC, 2005.

6 Charles Walaga, email to Peter Singer, April 2005. See also Paul Collier and Ritva Reinikka (eds), *Uganda's Recovery: The Role of Farms, Firms and Government*, Washington, DC: World Bank, 2001.

7 Charles Walaga, email to Peter Singer, April 2005.

8 Oxfam International, *Rigged Rules and Double Standards: Trade, Globalisation and the Fight Against Poverty*, Oxfam, 2002, pp. 10, 48, 53–55. www.maketradefair.com/assets/english/report_english.pdf. See also John Mellor, "Reducing Poverty, Buffering Economic Shocks—Agriculture and the Non-tradable Economy," in FAO Roles of Agriculture Project, *Expert Meeting Proceedings: First Expert Meeting on the Documentation and Measurement of the Roles of Agriculture, 19-21 March 2001*. FAO: Rome, 2001, p. 275. Available at ftp://ftp.fao.org/es/esa/roa/pdf/EMP-E.pdf. We owe this reference to Sophia Murphy, whose report *Agriculture Inc.* is currently in preparation for Oxfam America.

9 Brian Halweil, email to Peter Singer, February 2005.

10 For Kenya and Zimbabwe, see C. Dolan, J. Humphrey, and C. Harris-Pascal, "Value Chains and Upgrading: The Impact of U.K. Retailers on the Fresh Fruit and Vegetables Industry in Africa," Institute of Development Studies Working Paper 96, University of Sussex, 1988, as cited in R. Kaplinsky, "Spreading the Gains from Globalization: What Can Be Learned from Value-Chain Analysis?" *Problems of Economic Transition,* vol. 47 (2004), pp. 74–115. The analysis of South African peaches also comes from Raplinsky's article. The figure for banana workers is from FAO, *The State of Agricultural Commodity Markets: 2004*. FAO, Rome, 2004, p. 31.

11 Charles Walaga, email to Peter Singer, April 2005 and "The Development of the Organic Agriculture Sector in Africa: Potentials and Challenges," *Ecology and Farming*, no. 29, January-April, 2002.

12 Fairtrade Labelling Organisation International, www.fairtrade.net/sites/standards/general.html

13 TransFair USA, "Community Impacts," www.transfairusa.org/content/about/overview.php

14 Scientific Certified Systems, "Starbucks C.A.F.E. Practices" www.scscertified.com/csrpurchasing/starbucks.html

15 For background, see Sasha Courville, "Social Accountability Audits: Challenging or Defending Democratic Governance?" *Law and Policy*, vol. 25 (2003) pp. 269-297; for details of SA8000, see www.sa8000.org.

16 Michael Mitchell, email to Peter Singer, March 8, 2005.

17 International Institute of Tropical Agriculture, "Summary of Findings from the Child Labor Surveys in the Cocoa Sector of West Africa: Cameroon, Côte d'Ivoire, Ghana, and Nigeria," July 2002, www.iita.org/news/chlab-rpt.htm.

18 Reuters, Washington, "Lawmaker Shuns Valentine Candy, Cites Slavery Fear," February 10, 2005.

19 Oxfam International, *Rigged Rules and Double Standards: Trade, Globalisation and the Fight Against Poverty,* Oxfam, 2002, p. 55. www.maketradefair.com/assets/english/report_english.pdf; see also www.divinechocolate.com; Fairtrade Labelling Organizations International, "Fairtrade Standards for Cocoa for Small Farmers' Organizations," (December 2005) www.fairtrade.net/pdf/sp/english/Cocoa%20SP%20Dec%2005%20EN.pdf.

20 Brink Lindsey, *Grounds for Complaint? Understanding the "Coffee Crisis,"* www.freetrade.org/pubs/briefs/tbp-016.pdf.

21 R. H. Frank, T. Gilovich, and T. D. Regan, "Does studying economics inhibit cooperation?" *Journal of Economic Perspectives,* 7 (1993), pp. 159–171.

22 See, for example, Joseph Henrich et al, eds., *Foundations of Human Sociality: Economic Experiments and Ethnographic Evidence from Fifteen Small-Scale Societies,* Oxford University Press, New York, 2004; Colin Camerer, *Behavioral Game Theory: Experiments in Strategic Interaction*, Princeton University Press, Princeton, 2003.

23 See "CIW Anti-Slavery Campaign" www.ciw-online.org/slavery.html, and John Bowe, "'Nobodies:' Does Slavery Exist in America?" *The New Yorker,* April 21, 2003.

24 "2003 Robert F. Kennedy Human Rights Award Laureates," www.rfkmemorial.org/legacyinaction/2003_CIW.

25 Evelyn Nieves, "Fla Tomato Pickers Still Reap 'Harvest of Shame,'" *Washington Post,* February 28, 2005, www.washingtonpost.com/wp-dyn/articles/A58505-2005Feb27.html.

26 "Coalition of Immokalee Workers, Taco Bell, Reach Ground-Breaking Agreement," www.ciw-online.org/we%20won.html; Evelyn Nieves, "Accord with Tomato Pickers Ends Boycott of Taco Bell," *Washington Post*, March 9, 2005, www.washingtonpost.com/wp-dyn/articles/A18187-2005Mar8.html.

27 "Comments by Coalition of Immokalee Workers Co-Director Lucas Benitez at Press Conference, Announcing Settlement of the CIW's Taco Bell Boycott," March 8, 2005, www.ciw-online.org/lucasspeech.html.

## CHAPTER 12

1 Conversation with Jim Norwood, Director of Procurement, Meyer Natural Angus Beef, July 2005.

2 Additional sources used for this section are: John Mackey and Lauren Ornelas in discussion with Karen Dawn, "Watchdog," KPFK-FM, Los Angeles, May 3, 2004, reprinted as John Mackey, Karen Dawn, and Lauren Ornelas, "The CEO as Animal Activist: John Mackey and Whole Foods," in Peter Singer, ed., *In Defense of Animals,* Blackwell, Oxford, 2005. www.animalcompassionfoundation.org; www.wholefoodsmarket.com; Jon Gertner, "The Virtue in $6 Heirloom Tomatoes," *The New York Times Sunday Magazine,* June 6, 2004; Charles Fishman, "The Anarchist's Cookbook," *Fast Company,* July 2004; www.fastcompany.com/magazine/84/wholefoods.html; Stuart Truelson, "Whole Foods Markets and Animal Rights Groups Team Up," *The Voice of Agriculture,* January 31, 2005, www.fb.org/views/focus/fo2005/fo0131.html; Daniel McGinn, "The Green Machine," *Newsweek,* March 21, 2005, http://msnbc.msn.com/id/7130106/site/newsweek/; Seth Lubove, "Food Porn," *Forbes,* February 14, 2005, www.forbes.com/free_forbes/2005/0214/102.html; and personal communications with John Mackey.

3 John Dicker, *The United States of Wal-Mart*, Tarcher/Penguin, New York, 2005, p. 30.

## CHAPTER 13

1 Humane Research Council, Vegetarianism in the U.S.: A summary of quantitative research, August 2005, available on request from info@humaneresearch.org; Vegetarian Resource Group, "How many vegetarians are there?" www.vrg.org/journal/vj2003issue3/vj2003issue3poll.htm.

2 Thomas Frank, *What's the Matter with Kansas? How Conservatives Won the Heart of America*, Metropolitan Books, New York, 2004, p. 103.

3 Lantern Books, New York, 2000.

## CHAPTER 14

1 Organic Trade Association, drawing on various sources. See: www.ota.com/organic/mt/business.html; www.ota.com/organic/mt/consumer.html; and www.ota.com/organic/mt/food.html.

2 Reuters News Service, "U.K Organic Food Sales Top 1.1. Billion stg/yr," November 16, 2004, www.planetark.com/dailynewsstory.cfm/newsid/28166/newsDate/16-Nov-2004/story.htm

3 Erik Millstone and Tim Lang, *The Atlas of Food*, Earthscan, London, 2003, pp. 56–7.

4 Quoted from Michael Sligh and Caroline Christman, *Who Owns Organic? The Global Status, Prospects and Challenges of a Changing Organic Market*, Rural Advancement Foundation International—USA, Pittsboro, NC, 2003, p. 1.

5 Stephen Cadogan, "Babies Best Customers for Organic Food," *Irish Examiner*, February 17, 2005; Bernward Geier, "An Overview and Facts on Worldwide Organic Agriculture: Organic Trade a Growing Reality," ftp://ftp.fao.org/docrep/fao/006/ad429E/ad429E00.pdf, p. 9.

6 Brian Baker et al, "Pesticide residues in conventional, IPM-grown and organic foods: Insights from three U.S. data sets," *Food Additives and Contaminants*, Volume 19, No. 5, May 2002, pp. 427-446. A summary is available at: www.consumersunion.org/food/organicsumm.htm.

7 Cynthia L. Curl, Richard A. Fenske, and Kai Elgethun, "Organophosphorus Pesticide Exposure of Urban and Suburban Preschool Children with Organic and Conventional Diets," *Environmental Health Perspectives* vol. 111 (2003), pp. 377–382.

8 Sir John Krebs, "Is Organic Food Better For You?" Cheltenham Science Festival, June 5, 2003, www.food.gov.uk/news/newsarchive/2003/jun/cheltenham.

9 Michael Pollan, "Behind the Organic-Industrial Complex," *The New York Times*, May 13, 2001.

10 Organisation for Economic Co-Operation and Development, *Organic Agriculture: Sustainability, Markets and Policies*, CABI Publishing, Paris, 2003, p. 10. Available at www1.oecd.org/publications/e-book/5103071E.pdf.

11 D. Tillman, "The greening of the green revolution," *Nature*, vol. 396 (1998) pp. 211–2.

12 Mark Shepherd et al, "An Assessment of the Environmental Impacts of Organic Farming," A review for Defra-Funded Project OF0405, May 2003, pp. 26-34; available at www.defra.gov.uk/farm/organic/research/env-impacts2.pdf.

13 John Reganold et al, "Long-term effects of organic and conventional farming on soil erosion," *Nature*, vol. 330 (1987) pp. 370–372.

14 D. G. Hole et al, "Does Organic Farming Benefit Biodiversity?" *Biological Conservation*, vol. 122 (2005) pp. 113–130.

15 "Organic farms 'best for wildlife,'" BBC News, August 3, 2005, http://news.bbc.co.uk/2/hi/uk_news/4740609.stm.

16 D. Tillman, "The greening of the green revolution," *Nature*, 396 (1998) pp. 211–2.

17 Janet Larson, "Dead Zones Increasing in World's Coastal Areas," Earth Policy Institute, June 16, 2004, available at www.earth-policy.org/Updates/Update41.htm; Ron Brunoehler, "Resurrecting the Dead Zone," *The Corn and Soybean Digest*, May 1, 1998, available at www.cornandsoybeandigest.com/mag/soybean_resurrecting_dead_zone/.

18 Arnold Aspelin, *Pesticide Usage in the United States: Trends During the 20th Century*. CIPM Technical Bulletin 105, Center for Integrated Pest Management North Carolina State University, Raleigh, N. C., 2003, Pt V. www.pestmanagement.info/pesticide_history/five.pdf; T. Kiely, D. Donaldson, and A. Grube, *Pesticides Industry Sales and Usage: 2000 and 2001 Market Estimates*, Office of Pesticide Programs, US Environmental Protection Agency, 2004. www.epa.gov/oppbead1/pestsales/01pestsales/market_estimates2001.pdf.

19 Department of the Interior, U.S. Geological Survey, *The Quality of Our Nation's Waters: Nutrients and Pesticides*. USGS Circular 1225, Reston, VA, 1999. Available at http://water.usgs.gov/pubs/circ/circ1225/.

20 Mark Shepherd et al, "An Assessment of the Environmental Impacts of Organic Farming," A review for Defra-Funded Project OF0405, May 2003, available at www.defra.gov.uk/farm/organic/research/env-impacts2.pdf.

21 J. Pretty and A. Ball, "Agricultural Influences on Carbon Emissions and Sequestration: A Review of Evidence and the Emerging Trading Options," Centre for Environment and Society Occasional Paper 2001-03, University of Essex, 2001; see also J. Pretty et al, "The Role of Sustainable Agriculture and Renewable Resource Management in Reducing Greenhouse Gas Emissions and Increasing Sinks in China and India," in *Philosophical Transactions of the Royal Society (Series A: Mathematical, Physical and Engineering Sciences)*, vol. 360 (2002) pp. 1741–1761.

22 Paul Hepperly, "Organic Farming Sequesters Atmospheric Carbon and Nutrients in Soils," Rodale Institute, www.strauscom.com/rodale-whitepaper/.

23 On carbon sequestration and organic farming, see J. Pretty and A. Ball, "Agricultural Influences on Carbon Emissions and Sequestration: A Review of Evidence and the Emerging Trading Options," *Centre for Environment and Society Occasional Paper* 2001-03, University of Essex, 2001; Robert Jackson and William Schlesinger, "Curbing the U.S. Carbon Deficit," *Proceedings of the National Academy of Sciences of the United States*, vol. 101 (November 9, 2004) pp. 15827–15829, www.pnas.org_cgi_doi_10.1073_pnas.0403631101.

24 David Pimentel et al, "Environmental, Energetic, and Economic Comparisons of Organic and Conventional Farming Systems," *Bioscience*, vol. 55 (2005) pp. 573–582.

25 Mark Shepherd et al, "An Assessment of the Environmental Impacts of Organic Farming," A review for Defra-Funded Project OF0405, May 2003, pp. 49–53; available at www.defra.gov.uk/farm/organic/research/env-impacts2.pdf.

26 Mary Shelley, *Frankenstein*, London, first published 1818.

27 National Research Council, *Environmental Effects of Transgenic Plants*. Washington, DC: National Academy Press, 2002.

28 Robert H. Devlin, "Major factors influencing reliability of risk assessment data derived from laboratory-contained GH transgenic coho salmon," a paper presented to the 8th International Symposium on Risk Assessment of GMOs, and kindly made available to us by Dr Devlin; Rachel Borgatti and Eugene Buck, "Genetically Engineered Fish and Seafood," CRS Report for Congress, Congressional Research Service, The Library of Congress, December 7, 2004, www.ncseonline.org/nle/crsreports/04dec/RS21996.pdf.

29 Guelph Transgenic Pig Program, www.uoguelph.ca/enviropig/.

30 Food and Agriculture Organization of the United Nations, *The State of Food and Agriculture, 2003-4*. Rome, 2004, www.fao.org/documents/show_cdr.asp?url_file=/docrep/006/Y5160E/Y5160E00.HTM.

31 *ibid.*

32 Uma Lele, "Biotechnology: opportunities and challenges for developing countries," *American Journal of Agricultural Economics*, vol. 85, 2003, pp.1119–1125.

33 Food Standards Australia New Zealand, Report on the Review of Labelling of Genetically Modified Foods, December 2003, www.foodstandards.gov.au/_srcfiles/GM_label_REVIEW%20REPORT%20_Final%203_.pdf., Secs 9-10.

34 Associated Press, "Americans clueless about gene-altered foods," March 23, 2005, http://msnbc.msn.com/id/7277844/.

35 Lee Silver, *Challenging Nature: The Clash of Science and Spirituality at the New Frontiers of Life*, Ecco, New York, 2006.

36 Jacques Diouf, "Foreword" in Food and Agriculture Organization of the United Nations, *The State of Food and Agriculture, 2003-4*. Rome, 2004, p. viii. www.fao.org/documents/show_cdr.asp?url_file=/docrep/006/Y5160E/Y5160E00.HTM; G. J. Persley, "New genetics, food and agriculture: scientific discoveries—societal dilemmas," Paris, 2003, p. 8. www.icsu.org/2_resourcecentre/INIT_GMOrep_1.php4.

37 "Does Bt maize kill monarch butterflies," in Food and Agriculture Organization of the United Nations, *The State of Food and Agriculture, 2003-4*. Rome, 2004, Box 24, p. 71; see also John E. Losey, Linda S. Raynor, and Maureen E. Carter, "Transgenic pollen harms monarch larvae." *Nature*, vol. 399 (May 20, 1999), p. 214; Tom Clarke, "Monarchs safe from Bt," News@Nature.com, September 12, 2001.

38 Food and Agriculture Organization of the United Nations, *The State of Food and Agriculture, 2003-4*. Rome, 2004, p. 67. www.fao.org/documents/show_cdr.asp?url_file=/docrep/006/Y5160E/Y5160E00.HTM.

39 Andrew Pollack, "A Texas-Size Whodunit: On the Trail of Genetically Altered Corn from Azteca," *The New York Times*, September 30, 2000; "1999 Survey on Gene-Altered Corn Disclosed Some Improper Uses," Andrew Pollack, *The New York Times*, September 4, 2001.

40 Food and Agriculture Organization, *The State of Food Insecurity in the World*, 2000. Rome, 2000, p. 9.

41 Vaclav Smil, *Feeding the World: A Challenge for the Twenty-First Century*, MIT Press, Cambridge, MA, 2001, p. 315.

42 Julie Guthman, *Agrarian Dreams: The Paradox of Organic Farming in California*, University of California Press, Berkeley and Los Angeles, 2004, p. 169.

43 Jim Mason, interview with Elizabeth Henderson, February 22, 2005; Beth Holtzman, interview with Elizabeth Henderson, in Valerie Berton, ed., *The New American Farmer: Profiles of Agricultural Innovation*, U.S. Department of Agriculture Sustainable Agriculture Research and Education program, Beltsville, MD, 2001. pp. 65–67, www.sare.org/publications/naf/naf.pdf.

44 Rebecca Clarren, "Land of Milk and Honey," *Salon*, April 13, 2005, www.salon.com/news/feature/2005/04/13/milk/

45 Andrew Martin, "Organic Milk Debate," *Chicago Tribune*, January 10, 2005.

46 Steve Raabe, "Organic Farm Under Fire Over Pasture Rules," *Denver Post*, January 16, 2005.

47 Andrew Martin, "Organic Milk Debate," *Chicago Tribune*, January 10, 2005.

48 Michael Sligh and Caroline Christman, *Who Owns Organic? The Global Status, Prospects and Challenges of a Changing Organic Market*, Rural Advancement Foundation International—USA, Pittsboro, NC, 2003, p. 19.

49 Andrew Martin, "Panel Seeks to Put Organic Loophole Out to Pasture," *Chicago Tribune*, March 2, 2005.

50 Marion Nestle, "In Praise of the Organic Environment," *Global Agenda*, 2005, www.globalagendamagazine.com/2005/marionnestle.asp.

51 Center for Global Food Issues, www.cgfi.org.

## CHAPTER 15

1 Michelle Roberts, "Children 'harmed' by vegan diet," *BBC News*, February 21, 2005, http://news.bbc.co.uk/1/hi/health/4282257.stm; Jim McBeth, "Vegetarian diet 'bad for children,'" *The Scotsman*, February 22, 2005, http://thescotsman.scotsman.com/index.cfm?id=199842005.

2 A. R. Mangels and V. Messina, "Considerations in planning vegan diets: infants," *Journal of the American Dietetic Association*; 101 (2001) pp. 670-77; "Position of the American Dietetic Association and Dietitians of Canada: vegetarian diets." *Canadian Journal of Dietetic Association Practice and Research*, 64 (2003), pp. 62-81, available at www.ncbi.nlm.nih.gov/entrez/query.fcgi?cmd=Retrieve&db=PubMed&list_uids=12826028&dopt=Citation.

3 Pramil Singh, Joan Sabaté, and Gary Fraser, "Does low meat consumption increase life expectancy in humans?" *American Journal of Clinical Nutrition*, 2003;78(suppl):526S–32S.

4 www.vegsource.com/articles2/ncbs_vegan_study.htm

5 "Supplement: Animal Source Foods to Improve Micronutrient Nutrition in Developing Countries," *Journal of Nutrition*, November 2003, vol. 133, pp. 3875s-4054s; for funding from the National Cattleman's Beef Association, see Charlotte Neuman et al, "Animal Source Foods Improve Dietary Quality, Micronutrient Status, Growth and Cognitive Function in Kenyan School Children: Background, Study Design and Baseline Findings" in the same supplement, pp. 3941s-3949s, fn. 2.

6 See www.veganoutreach.org. Jack Norris's essay "Staying Healthy on Plant-Based Diets," available from a link on this Web site or at www.veganhealth.org/sh/, is an excellent practical guide to what people contemplating vegetarian and vegan diets should know about nutrition.

7 "Position of the American Dietetic Association and Dietitians of Canada: Vegetarian diets," *Journal of the American Dietetic Association*, vol. 103 (2003), p. 749.

8 Lindsay Allen, "Interventions for Micronutrient Deficiency Control in Developing Countries: Past, Present and Future," *Journal of Nutrition*, vol. 133 (November 2003) Supplement, pp. 3875S-3878S; Colin Tudge, *So Shall We Reap*, Allen Lane, 2003, p. 125.

9 Jonathan H. Siekmann, Lindsay H. Allen, et al, "Kenyan School Children Have Multiple Micronutrient Deficiencies, but Increased Plasma Vitamin B-12 Is the Only Detectable Micronutrient Response to Meat or Milk Supplementation." *Journal of Nutrition*, vol. 133 (November 2003) Supplement, pp. 3972S–3980S.

10 Colin Tudge, *So Shall Ye Reap*, Allen Lane, London, 2003, pp. 334–35.

11 Immanuel Kant, *The Moral Law: Kant's Groundwork of the Metaphysic of Morals*, translated by H. J. Paton, Hutchinson University Library, London, 1966 (first published 1785), p. 67.

12 Andrew Tyler, email, June 2005; Joyce D'Silva in conversation, London, July 2005.

13 Kristin Dizon, "Seattle man amazes everyone in 135-mile marathon—including himself," *Seattle Post-Intelligencer*, July 22, 2005, http://seattlepi.nwsource.com/othersports/233630_jurek22.html; Carl Lewis, "Introduction" in Jannequin Bennett, *Very Vegetarian*, Rutledge Hill Press, Nashville, TN, 2001.

## CHAPTER 16

1 Frances Moore Lappé, *Diet for a Small Planet*, Ballantine, New York, 1971.

2 Erik Marcus, *Meat Market: Animals, Ethics, and Money*, Brio Press, Ithaca, NY, 2005, pp. 255, citing W. O. Herring and J. K. Bertrand, "Multi-trait prediction of feed conversion in feedlot cattle," *Proceedings from the 34th Annual Beef Improvement Federation Annual Meeting*, Omaha, NE, July 10-13, 2002, www.bifconference.com/bif2002/BIFsymposium_pdfs/Herring_02BIF.pdf.

3 Erik Marcus, *Meat Market: Animals, Ethics, and Money*, Brio Press, Ithaca, NY, 2005, pp. 256, citing Pork Facts, 2001/2002, National Pork Board, Des Moines, Iowa.

4 Tyson Foods Inc, 2004-5 Investor Fact Book, p. 5, http://media.corporate-ir.net/media_files/irol/65/65476/reports/04_05_factbook.pdf, citing figures from the National Chicken Council; Erik Marcus, *Meat Market: Animals, Ethics, and Money*, Brio Press, Ithaca, NY, 2005, pp. 255–56, citing Glen Fukomoto and John Replogle, *Livestock Management*, Cooperative Extension Service, College of Tropical Agriculture and Human Resources, University of Hawaii, Manoa, April 1999, and F. H. Ricard, "Carcass Conformation of Poultry and Game Birds," *Proceedings of the 25th World's Poultry Science Association Symposium on Meat Quality in Poultry and Game Birds*, Norwich, 1979, pp. 31–5.

5 G. Sarwar and F. McDonough, "Evaluation of protein digestibility-corrected amino acid score method for assessing protein quality of foods," *Journal of the Association of Official Analytical Chemists*, vol. 73 (1990), pp. 347–56; Food and Agriculture Organization/World Health Organization *Protein Quality Evaluation: Report of the Joint FAO/WHO Expert Consultation*, FAO Food and Nutrition paper 51, FAO, Rome, 1991.

6 Vaclav Smil, *Feeding the World: A Challenge for the Twenty-First Century*, MIT Press, Cambridge, MA, 2000, p. 145; Vaclav Smil, "Eating Meat: Evolution, Patterns, and Consequences," *Population Development Review*, vol. 28 (2002) p. 619.

7 Keite Camacho, "Brazil's Deforestation Worries Scientists," *Brazzil Magazine*, July 1, 2004, www.brazzil.com/content/view/2005/.

8 Gaverick Matheny and Kai Chan, "Human Diets and Animal Welfare: The Illogic of the Larder," *Journal of Agricultural and Environmental Ethics*, vol. 18 (2005), pp. 579–94.

9 *Newsweek*, February 22, 1981.

10 J. L. Beckett and J. W. Oltjen, "Estimation of the Water Requirements for Beef Production in the United States," *Journal of Animal Science*, vol. 71 (1993) pp. 818–826.

11 D. Pimentel et al, "Water resources: agriculture, the environment, and Society," *BioScience*, vol. 47 (1997), pp. 97–106. For simplicity of comparison, throughout this section we have converted metric figures to pounds and U.S. gallons.

12 D. Pimentel et al, "Water Resources: Agricultural and Environmental Issues," *BioScience*, vol. 54 (2004), pp. 909–918.

13 A. K. Chapagain and A. Y. Hoekstra, *Water Footprints of Nations: Volume 1: Main Report*, Unesco-IHE Institute of Water Education, Delft, November 2004, Table 4.1, p. 41.

14 USS Maddox Destroyer Association, www.ussmaddox.org/.

15 A. K. Chapagain and A. Y. Hoekstra, *Water Footprints of Nations: Volume 1: Main Report*, Unesco-IHE Institute of Water Education, Delft, November 2004, Table 4.2, p. 42.

16 Edward Abbey, "The Cowboy and His Cow," a speech given in Missoula, Montana in April 1985, reprinted in George Wuerthner and Mollie Matteson, eds., *Welfare Ranching: The Subsidized Destruction of the American West*, Island Press, Washington, 2002, p. 60.

17 World Resources Institute, *World Resources, 1998-99: A Guide to the Global Environment*, Washington D.C., 1998, p. 157.

18 George Wuerthner and Mollie Matteson, eds., *Welfare Ranching: The Subsidized Destruction of the American West*, Island Press, Washington, 2002.

19 *Welfare Ranching*, p. xiii.

20 Jack Rosenberger, "Wasting the West," *E Magazine*, July/August 2004, www.emagazine.com/view/?1855.

21 Australian Government, Department of the Environment and Heritage, "2005 Commerical Kangaroo Harvest Quotas," December 2004, www.deh.gov.au/biodiversity/trade-use/publications/kangaroo/quotas-background-2005.html.

22 Christie Aschwanden, "Learning to Live With Prairie Dogs," *National Wildlife* vol. 39, no. 2, April/May 2001, www.nwf.org/nationalwildlife/article.cfm?issueID=34&articleID=327.

23 U.S. Department of Agriculture, Wildlife Services, Table 10T "Number of Animals Killed and Methods Used by the WS Program, FY2004" www.aphis.usda.gov/ws/tables/04tables.html.

24 "It's better to green your diet than your car," *New Scientist*, 17, December 2005, p. 19, www.newscientist.com/channel/earth/mg18825304.800.

25 The Editors, "Meat: Now, it's not personal," *World Watch Magazine*, July/August 2004, www.worldwatch.org/pubs/mag/2004/174/.

## CHAPTER 17

1 Hugh Fearnley-Whittingstall, *The River Cottage Meat Book*, Hodder and Stoughton, London, 2004, p. 24.

2 Michael Pollan, "An Animal's Place," *The New York Times Sunday Magazine*, November 10, 2002; see also Michael Pollan, *The Omnivore's Dilemma: A Natural History of Four Meals*, Penguin, New York, 2006.

3 Sholto Byrnes, "Roger Scruton: The Patron Saint of Lost Causes," *The Independent*, July 3, 2005, http://enjoyment.independent.co.uk/books/features/article296509.ece

4 Roger Scruton, *Animal Rights and Wrongs*, 3rd ed., Claridge Press, 2003.

5 Matthew Scully, "Fear Factories: The Case for Compassionate Conservatism—for Animals," *The American Conservative*, May 23, 2005; George F. Will, "What We Owe What We Eat," *Newsweek*, July 18, 2005; Matthew Scully, *Dominion: The Power of Man, the Suffering of Animals, and the Call to Mercy*, St Martin's Press, New York, 2003.

6 Joseph Ratzinger, *God and the World: Believing and Living in Our Time. A Conversation with Peter Seewald*. San Francisco: St.Ignatius Press, 2002, p. 78; for other Christian views that point in the same direction, see Matthew Scully, *Dominion: The Power of Man, the Suffering of Animals, and the Call to Mercy*, St Martin's Press, New York, 2002; and Andrew Linzey, *Animal Theology*, University of Illinois Press, Chicago, 1994.

7 See, for example, Peter Carruthers, *The Animals Issue: Moral Theory in Practice*, Cambridge University Press, Cambridge, 1992.

8 Benjamin Franklin, *Autobiography*, New York, Modern Library, 1950, p. 41.

9 T. Colin Campbell and Thomas Campbell, *The China Study: The Most Comprehensive Study of Nutrition Ever Conducted and the Startling Implications for Diet, Weight Loss and Long-Term Health*, Benbella, Dallas, TX, 2005.

10 Jonathan Swift, *A Modest Proposal for Preventing the Children of Poor People from Being a Burthen to Their Parents or Country, and for Making Them Beneficial to the Public*, first published 1729, reprinted in Tom Regan and Peter Singer, eds., *Animal Rights and Human Obligations*, Prentice-Hall, Englewood Cliffs, NJ, 1976, pp. 234–237.

11 For a powerful argument for this position, see Paola Cavalieri, *The Animal Question: Why Non-Human Animals Deserve Human Rights*. Tr. Catherine Woollard, Oxford University Press, New York, 2001.

12 Stephen Budiansky, *The Covenant of the Wild*, HarperCollins, New York, 1992.

13 Hugh Fearnley-Whittingstall, *The River Cottage Meat Book*, Hodder and Stoughton, London, 2004, pp. 23–25.

14 Henry Salt, "The Logic of the Larder," first published in Henry Salt, *The Humanities of Diet*, The Vegetarian Society, Manchester, 1914, reprinted in Tom Regan and Peter Singer, *Animal Rights and Human Obligations*, Prentice-Hall, Englewood Cliffs, NJ, 1976, p. 186.

15 See Derek Parfit, *Reasons and Persons*, Clarendon Press, Oxford, 1984, Part IV.

16 Roger Scruton, "The Conscientious Carnivore" in Steve Sapontzis, ed., *Food For Thought: The Debate over Eating Meat*, Prometheus, Amherst, NY, 2004, p. 88.

17 Gaverick Matheny and Kai Chan, "Human Diets and Animal Welfare: The Illogic of the Larder," *Journal of Agricultural and Environmental Ethics*, vol. 18 (2005), pp. 579–94; and personal correspondence with Gaverick Matheny, April 2005.

18 Steven Davis, "The Least Harm Principle May Require that Humans Consume A Diet Containing Large Herbivores, Not A Vegan Diet," *Journal of Agricultural and Environmental Ethics*, vol. 16 (2003) pp. 387–394.

19 Gaverick Matheny, "Least Harm: A Defense of Vegetarianism from Steven Davis's Omnivorous Proposal," *Journal of Agricultural and Environmental Ethics*, vol. 16 (2003), pp. 505–511.

20 Todd Purdum, "High Priest of the Pasture," *New York Times Style Magazine*, Living, Spring 2005, pp. 76–79. The comment from Daniel Salatin about his father is taken from the 13th Annual Wisconsin Grazing Conference, February 14, 2005, www.grassworks.org/Conference/conference.htm.

21 Joel Salatin, "Family Friendly Farming," *AcresUSA*, June 2000, available at www.acresusa.com/toolbox/reprints/familyfriendly_jun00.pdf.

22 Anne Fanatico, *Sustainable Poultry: Production Overview*, ATTRA—National Sustainable Agriculture Information Service, March 2002, http://attra.ncat.org/attra-pub/poultryoverview.html.

23 See Herman Beck-Chenoweth, *Free-Range Poultry Production and Marketing*, Back40Books, Hartshorn, Missouri, 2001. The quote is taken from the same author's "Free Range, Pastured Poultry, Chicken Tractor—What's the Difference?" www.free-rangepoultry.com/compare.htm.

24 George Devault, "'Chicken Day' at the Farm of Many Faces," *The New Farm*, August 2002, www.newfarm.org/features/0802/chicken%20day/print.html.

25 The percentage disapproving of hunting for food varies, but is usually at least 12 percent and in some state polls has been as high as 34 percent. See Mark Damian Duda and Kira C. Young, "American Attitudes Toward Scientific Wildlife Management and Use of Fish and Wildlife: Implications for Effective Public Relations and Communications Strategies," *Transactions of the 63rd North American Wildlife and Natural Resources Conference*, 1998, pp. 589–603, www.responsivemanagement.com/download/reports/AmericanAttitudes.pdf.

26 *River Cottage Meat Book*, p. 153.

27 Frederick Pohl and C. M. Kornbluth, *The Space Merchants*, Ballantine, New York, 1952.

28 Winston Churchill, *Thoughts and Adventures*, Thornton Butterworth, London, 1932, pp. 24–27.

29 For more information on progress in making cultured meat, see www.new-harvest.org.

30 www.meetup.com, visited January 30, 2006.

31 Anonymous, "Why Freegan?" Food Not Bombs Houston, December 30, 2002, http://fnbhouston.org/20021230-2815.html.

32 Our account of freeganism draws on Web sites like http://freegan.info/ and the writings of Adam Weissman.

33 Lance Gay, "Food Waste Costing Economy $100 billion, Study Finds," Scripps Howard News Service, August 10, 2005, www.knoxstudio.com/shns/story.cfm?pk=GARBAGE-08-10-05&cat=AN

## CHAPTER 18

1 We checked prices at Wal-Mart and other major supermarket chains, and calculated that the protein content of dried lentils, beans, and peas cost between half a cent and 1.5 cents, per gram. One gram of protein from peanut butter cost about 2 cents. Frozen edamame soybeans and lima beans yielded protein at 4 and 6 cents per gram, respectively. Tofu worked out at 6 cents per gram. The protein content in chicken cost, depending on the form of chicken bought, between 3 and 9 cents per gram. Vegan meat alternatives like soyburgers and chicken-like patties came to between 6 and 11 cents per gram of protein. Textured vegetable protein from www.healthy-eating.com costs about 1 cent per gram of protein. (Prices were checked in August 2005)

2 Judy Putnam, "US Food Supply Providing More Food and Calories," *FoodReview*, vol. 22, no. 3 (September 1999), Table 1, p. 6; www.ers.usda.gov/publications/foodreview/sep1999/frsept99a.pdf.

3 Javachandran Variyam, "The Price is Right: Economics and Obesity," *Amber Waves*, February 2005, www.ers.usda.gov/AmberWaves/February05/Features/ThePriceIsRight.htm; James Meikle, "Obesity: rising fears of cancer time bomb," the *Guardian*, January 8, 2004, http://society.guardian.co.uk/publichealth/story/0,,1118252,00.html.

4 Kenneth E. Thorpe et al., "The Rising Prevalence Of Treated Disease: Effects On Private Health Insurance Spending," *Health Affairs,* vol. 10, June 27, 2005, online at http://content.healthaffairs.org/cgi/content/abstract/hlthaff.w5.317.

5 U.S. Department of Health and Human Services, Centers for Disease Control, "Overweight and Obesity: Economic Consequences," 2005, www.cdc.gov/nccdphp/dnpa/obesity/economic_consequences.htm

6 Amanda Paulson, "One woman's quest to enjoy her dinner without guilt," *Christian Science Monitor*, October 27, 2004, http://csmonitor.com/2004/1027/p15s02-lifo.htm.

7 According to figures for 2003 from the Food and Agriculture Organisation of the United Nations, http://faostat.fao.org/faostat/collections?subset=agriculture.

# INDEX

ALSO AVAILABLE IN ARROW

# *No god But God*

## Reza Aslan

**'Reza Aslan's *No god But God* is just the history of Islam I needed, judicious and truly illuminating.' A.S. Byatt – *Guardian* Books of the Year 2005**

Though it is the fastest growing religion in the world, Islam remains shrouded in ignorance and fear. What is the essence of this ancient faith? Is it a religion of peace or of war? How does Allah differ from the God of Jews and Christians? Can an Islamic state be founded on democratic values such as pluralism and human rights?

In No god but God, challenging the 'clash of civilisations' mentality that has distorted out view of Islam, Aslan explains this faith in all its complexity, beauty and compassion.

A revelation, an opening up of knowledge too long buried, denied and corrupted by generations of men . . . Muslim keepers of the latter will rage against Reza Aslan as his careful scholarship and precise language dismantles their false claims and commands . . . Acutely perceptive.'
*Independent*

'Aslan . . . is a superb narrator, bringing each century to life with vivid details and present tense narration that make popular history so enthralling . . . Illuminating . . . A terrific read'
Glasgow Herald

arrow books

ALSO AVAILABLE IN ARROW

# *The Road Taken*

## Michael Buerk

**'Dawn, and as the sun breaks through the piercing chill of night on the plain outside Korem it lights up a biblical famine, now, in the Twentieth Century.'**

Those words opened Michael Buerk's first report on the Ethiopian famine for the 6 o'clock news on October 24th 1984. His reports sent shock waves round the world. The Live Aid concert, a direct consequence of Bob Geldof watching that broadcast, was watched by half the planet.

Michael Buerk has reported on some of the biggest stories in our lifetime: the Flixborough chemical plant fire, the Birmingham pub bombing, Lockerbie. He was in Buenos Aires at the start of the Falklands War; he reported the death throes of apartheid in South Africa.

He was the face of the BBC flagship evening news for many years and has fronted everything from the popular BBC1 series 999 to the erudite Radio 4 programme *The Moral Maze*. He has won every major award and is universally admired and respected for his intelligent and honest journalism. Here, he also reveals the private Michael Buerk, his bigamist father, his long and happy marriage to Christine and his delight at fatherhood.

'An exceedingly good book'
Michael Parkinson

arrow books

arrow books